W0235383

A Professional Guide for Students, Recent Graduates, and Beyond

Building Your Career in STEM

Professional Development Guides

Series editors:
Rowan Brookes, *Bendelta, Australia*
Christopher Thompson, *Monash University, Australia*
Samantha Pugh, *Leeds University, UK*

For a list of titles in this series see:
rsc.li/profdevgu

How to obtain future titles on publication:
A standing order plan is available for this series. A standing order will bring delivery of each new volume immediately on publication.

For further information please contact:
Book Sales Department, Royal Society of Chemistry, Thomas Graham House, Science Park, Milton Road, Cambridge, CB4 0WF, UK
Telephone: +44 (0)1223 420066, Fax: +44 (0)1223 420247
Email: booksales@rsc.org
Visit our website at books.rsc.org

Building Your Career in STEM

A Professional Guide for Students, Recent Graduates, and Beyond

Edited by

Angela Ziebell
Deakin University, Australia
Email: a.ziebell@deakin.edu.au

and

Rebecca Yee
Transition Energy Training Pty Ltd., Australia
Email: rsl.yee@gmail.com

ROYAL SOCIETY
OF **CHEMISTRY**

Professional Development Guides No. 1

Print ISBN: 978-1-83916-711-9
PDF ISBN: 978-1-83767-218-9
EPUB ISBN: 978-1-83767-219-6
Print ISSN: 2752-7794
Electronic ISSN: 2752-7808

A catalogue record for this book is available from the British Library

© The Royal Society of Chemistry 2024

All rights reserved

Apart from fair dealing for the purposes of research for non-commercial purposes or for private study, criticism or review, as permitted under the Copyright, Designs and Patents Act 1988 and the Copyright and Related Rights Regulations 2003, this publication may not be reproduced, stored or transmitted, in any form or by any means, without the prior permission in writing of The Royal Society of Chemistry or the copyright owner, or in the case of reproduction in accordance with the terms of licences issued by the Copyright Licensing Agency in the UK, or in accordance with the terms of the licences issued by the appropriate Reproduction Rights Organization outside the UK. Enquiries concerning reproduction outside the terms stated here should be sent to The Royal Society of Chemistry at the address printed on this page.

Whilst this material has been produced with all due care, The Royal Society of Chemistry cannot be held responsible or liable for its accuracy and completeness, nor for any consequences arising from any errors or the use of the information contained in this publication. The publication of advertisements does not constitute any endorsement by The Royal Society of Chemistry or Authors of any products advertised. The views and opinions advanced by contributors do not necessarily reflect those of The Royal Society of Chemistry which shall not be liable for any resulting loss or damage arising as a result of reliance upon this material.

The Royal Society of Chemistry is a charity, registered in England and Wales, Number 207890, and a company incorporated in England by Royal Charter (Registered No. RC000524), registered office: Burlington House, Piccadilly, London W1J 0BA, UK, Telephone: +44 (0) 20 7437 8656.

For further information see our website at www.rsc.org

Printed in the United Kingdom by CPI Group (UK) Ltd, Croydon, CR0 4YY, UK

Foreword

When I was an undergraduate student, all I needed to worry about was mastering the chemistry knowledge and related laboratory skills. It was this knowledge and these skills that would contribute to my final grade and that grade is what employers would largely base their selection processes on. These were simpler times!

For most of the time that I have taught chemistry in universities, there has been a growing focus on teaching—and assessing—transferable skills to STEM students of all levels. These skills started out as a relatively modest list: group work, problem solving, and communication. The length of that list has grown over the decades to include employability, commercial awareness, cultural awareness, entrepreneurship, creativity, global citizenship, and many more. This situation has been driven by employers who look to universities to fulfill their needs in the form of well-rounded, capable graduates who can hit the ground running and make a worthwhile contribution to their workplace.

Teaching discipline-specific content is fairly straightforward, especially for academics with expert subject knowledge. Teaching content whilst also infusing opportunities to develop a broad set of skills is much more challenging and such provision is patchy across higher education. It is also the case that many students, and even recent graduates, remain largely ignorant of the need for them to develop these skills and ignorant also of the reasons why academics persist in using teaching approaches that help to develop these skills.

This book is a timely addition as it hands the power for developing transferable skills to students and graduates themselves. Without this book, skills development is something that is *done to* them rather than something that they can *own, enhance, exploit,* and *use to their advantage* in the future. The authors tackle not just transferable skills development—where academic interest often ends—but extends the scope to cover the complete range of professional skills. These skills really do allow students and fresh graduates to take control of their professional destiny; targeting what they want to do, preparing for specific roles, building long-term networks and support systems, and ultimately making a meaningful contribution to their workplace, whatever and wherever that may be. The focus of the book is not what employers, or society, needs, but what the students and graduates need to get to where they plan to be. The focus on the readers themselves, rather than on educational approaches, is refreshing. It will benefit future students, scientists, and STEM graduates directly and, ultimately, their employers too.

Professor Tina Overton
Hull

Series Preface

Unlock your full potential as a STEM professional with this comprehensive series of books on professional development and employability skills. From building a standout online profile to navigating the workplace and advancing your career, our expert authors provide practical guidance and insightful strategies for success. Whether you're a recent graduate or an early career professional, these books will equip you with the tools and knowledge you need to excel in today's competitive job market and achieve your professional goals.

Preface

This book is designed to help STEM students and graduates (recent and not so recent) to build their career in professions where they use their STEM training. The world generally has a specific idea of what a STEM career looks like (white coats, people coding, people in high-vis clothing with large equipment, *etc.*). In reality, people who have studied STEM subjects go on to roles in all sectors and work in all sorts of environments. This can make navigating a STEM career effectively more difficult than professions where there is a straight line between training and workplace (*e.g.*, nursing, teaching, architecture, *etc.*). A student or graduate has more options. This is a great thing, but requires more decisions and maybe more planning.

This book grew out of the recognition that, while there are some great career subjects at many universities, not everyone had access to those subjects. That could be because their university doesn't offer a career unit, they had no room in their course, they just didn't know about the value of such a unit until too late in their studies, or possibly the offering was extracurricular with the student needing to work to support themselves rather than being able to do an optional extra class. With this book, students and graduates can work to build their career independent of timing or university offerings. The authors are all passionate about providing a resource that is easy to work through, supportive and useable by a large range of students. The authors are also very conscious that they all wish they could have had this sort of resource when they were students or graduates. While everyone will make the odd career development mistake, some of the stories in the book where the authors made mistakes would have been prevented by a book like this.

The authors would like to thank the many dozens of interviewees that they talked to during the preparation of this book. Your thoughts are a priceless addition, helping the reader to gain further insight into your specific workplace or situation. Please know that even if you don't see a lot of quotes from your interviews in the book (we wanted to feature some from everyone), each interview helped inform and develop the book. The authors would also like to thank the student groups from Deakin University who helped collect interview data in 2022 and 2023. The work that you did processing and transcribing the interviews meant that we could have representation from more people, countries and sectors. You will note that location is not listed with the quotes. This is for a few reasons. Firstly, the insights from these quotes are not dependent on the country people were in when they had the experience. This book is designed more generally that that. Secondly, many of our interviewees have live/worked in multiple countries, so listing one country wouldn't have made a lot of sense. We hope that through the interview quotes, we have provided a little window into the world of work that helped you build your career in STEM too.

Angela Ziebell
Rebecca Yee

Table of Contents

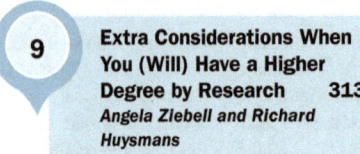

Introduction

ANGELA ZIEBELL AND MICHELLE HILL

Deakin University, Australia

A Tool for Navigating Your Career Development

The availability of good career advice for STEM students is highly dependent on where you go to school (both secondary and university) and how actively you feel, or felt, you can pursue that advice. Not everyone loves going to career days, and not everyone is great at navigating institutional systems to find help. But everyone deserves good career advice and help that will enable them to work out what they want from their career and how to navigate forward. We hope this book allows STEM people who are early in their career to navigate quickly to a job that is a great fit for them.

Everyone's situation is different, though. So in this book, instead of telling you what to do, we help you understand the employment environment that you are already within or that you are headed to. When we say work "environment", this is a combination of the influences from your discipline, where in the world you want to work and the industry you are considering. All these come together to form what we can call workplace culture. It is strongly influenced by the culture of the country the organisation is set in, but there is much more than that. Some environments are more or less causal, some will give you more independence, some will give you more training, others need a lot of technical skills, or you might need more transferrable skills. We will help you work through all these considerations and many, many more. At the end, you will understand the environment (or environments) you might work in.

But we will not just talk about work environments in this book. The book starts by helping you understand how to think about different types of skills you have, how to develop them, and, very importantly, how to articulate them so others can understand what you are capable of. The first specific skill we cover is commercial awareness. Commercial awareness isn't normally an obvious part of STEM education, although courses vary a lot. We will talk about why commercial awareness is important for everyone, even if your role is not directly commercial. Then we will look at networking, a key aspect of being connected to other professionals.

Figure 1.1   Map of this book.

Networking often has a bad reputation. We will take you through how to work out how to network in a way that works for you (Figure 1.1).

Chapter 5: Portfolio Building gives you the tools to start developing your career right from the first year of university. Portfolios are still applicable as you graduate and even after graduation. Portfolios can help you keep track of what you have achieved and plan what you want to do next and how. Next, we have Chapter 6: Resumes and Cover Letters and Chapter 7: Addressing Key Selection Criteria and Interviews. Chapters 6 and 7 help very directly with how you prepare to apply for a job, how you write your documents and how to get ready for an interview once your application gets you to the next step.

Finally, we look at the additional considerations when you have completed a research master's or a PhD. How does the process change for you? Or how might it change for you if you are thinking of doing a research degree? If you are in this group, all the other chapters are just as applicable as for people without a research degree. But in building the book, we wanted to specifically cover the process of getting a job after a research degree.

We Help You Find the Answers That Are Not in This Book

Because we aren't magic, we don't know where you will be interested in working and the exact question you might have about a specific company or workplace norm. So we try to help guide you through how to get advice from local sources that do know that information. Or, at the very least, these sources can help you work out how to find it out.

Using This Book – Based on Career Development Stage

Before we look at how this book can be used, please understand that this is a resource to help your career development and there is no one way to either progress through the book or to engage in any of the activities. There is no one way to progress your career. Every career is unique and development of your career should take in what you want, what your skills and attributes are, how you can develop them further, where you want to work, what qualifications you have (or are getting), and what your financial and family circumstances are. And then, at every life stage (about 5–10 years), any or all of those points may change.

The saying "everyone is different" couldn't be truer when it comes to careers and what you want out of your career. The next sections provide an orientation to how you might use the book based on the stage you are at within your student-professional journey. But you may be more or less progressed in a certain area. You might find a particular area challenging or already have experience with it. Move through the book purposefully, but in your own way.

Hint: Every career is different because everyone is different.

Students early in degrees can use this book to plan out and better understand future experiences. For example, you might become interested in completing an internship, which can take time to plan, you might get a different part-time job to build a skill or decide that adding a different minor to your degree would help you expand a desired technical or transferable skill.

Chapter 5: Portfolio Building is particularly useful for new students as it guides students through building experiences inside and outside your degree so that when you reach the end of your degree, you have a collection of experiences that have helped build you as a professional. This means you have generally built a sense of what you might want, and you have built some experience that helped you understand the workplace and build your resume.

By understanding concepts like transferrable skills and the need for commercial awareness early, you will be able to better understand some learning experiences your educators give you. You will see "the point" where others might not. For example, group projects might not just be a burden, they might be an occasion to develop organisational and leadership skills and have an interesting anecdote for your resume in a couple of years (Figure 1.2).

Figure 1.2   You spend years in classes. How many of them have you thinking about your employability or your career path? Photo from Desola Lanre-Ologun on Unsplash.

"Advice I would you give to someone just starting in their STEM degree is to just dabble probably a little bit of everything. Okay, you never know what you might like and what pathway you might take."

– Ruth, science graduate working in the retail food industry as a line manager, former business owner.

"Starting the degree? I think it's important to recognise that you should be trying to get more out of your science degree than just hardcore science, that you should be using it as an opportunity to use science for business and economics and arts. It's intertwined, and if you can't get that in your degree, then try and pick up electives. Supplement it in ways that you wouldn't expect. And for advice towards the end of the degree, I think it would be that you should trust what you have done is enough and you have a unique experience that only other science students have and you'll be fine."

– Megan, science graduate with honours in science education. Started in a consulting graduate program, moved to policy in government.

Higher year level students can still use all parts of the book but will likely want to make sure they get up to speed on Chapter 2: Transferrable Skills and Reflection, Chapter 4: Networking and Chapter 5: Portfolio Building a little more rapidly so they have time to act on any realisations before graduation. At this point in your studies, you have possibly had some experience that you could use to start your portfolio, therefore it's not just a case of planning future learning experiences.

You can spend time reflecting and working out what you have learnt so far, how you learnt it, what it means to you, and why. Understanding these points can help you direct the rest of your studies. For example, fine tune your elective choice, select a particular project topic around something you want to learn more about, or find a job that adds or expands on a skill. If it does not make sense how this will help, it is all explained in Chapters 2 and 5.

As you read and go through the activities, you will tend to see "the point" in more of the activities and learning that you are being given in class. It is easier to appreciate the skills you are being taught once you understand how those skills work and why they are important in career development.

Returning students or those redirecting their career will likely come in with more background knowledge than new students due to prior work experience. This book can be useful in a few ways. Many of us go on to a career without learning deeply about what we want from our career and how to best learn, develop ourselves, and communicate our skills and achievements to the world. That was certainly the experience of all the authors of this book. Now that you have had some career experience, you have a lot of skills and achievements that you can work on articulating. Those same experiences mean that you might have lots of thoughts about careers but may not have ways of moving forward in your thinking, working out what you want, why you want it, and how to make a plan for moving forward.

By working through this book, you can develop as a professional and gain a deeper level of understanding of where you might have knowledge gaps about how to manage your career. You might find that in some places you will skim through aspects that you are already comfortable with and feel a sense of achievement around. Hopefully, when this happens, you can take a moment to appreciate how far you have come since you were an undergraduate. In other areas, there might be more to work through. For example, you have more to communicate when applying for a job. This is a good thing, but it can take time to work methodically through what you are including in which job applications and why.

If you are also a returning student, you have the advantage of having professional experiences to reflect on and learn from, but you also have the advantage of being able to work out what you want to get from your current studies and how you will get it. You can use this book to work through both of those processes while building your general skills around networking, commercial awareness, and the job application and interview process. Again, some people in this position might not need some aspects of these chapters as much. When you get to a section that you are already comfortable with, it can still be good to read it to see if you can further improve. It is also worthwhile taking a moment to reflect on how much you have learnt since you were last a student and what that means for you and why.

Recent graduates and people already working can immediately use this book in their day-to-day life and will have examples that can be used in the activities most weeks, if not every week. To new graduates, I hope that working through our job application related Chapters 6 and 7 can help you get on top of any fears you might have about getting a good

Figure 1.3   Recent graduates have an intense and sometimes stressful journey working through what they want with their career. Photo from Juan Ramos on Unsplash.

professional resume and cover letter together for a job application. If you are pressed for time, it would be very fair if you find yourself using just the application-related chapters initially until you find your first role. Then, with a first role worked out, you can come back and branch out to further developing your career using the other chapters.

Many young employees got their first job without having much of a plan or maybe not a lot of skill navigating the job "hunting" process. If this is you (and it was certainly me), this is your chance to observe while you are at work and improve your skills while working through the book. While being gainfully employed, you can explore what might be the next opportunity, whether that's with your current employer or in a completely different field (Figure 1.3).

Graduate (research) students come in many forms: you might be a returning student or a continuing student. You might be extending your knowledge or redirecting your focus. For returning students, see the previous section too.

All STEM students and graduates can learn from any of the chapters. But, as we talk about in detail in Chapter 9: Higher Research Degree Considerations, those that do or have done a research degree tend to have a number of additional considerations. It's rare to find this discussion in career books, so we thought it especially important to include it in our book. As a higher research degree (HDR) student (either master's or PhD), there are a lot fewer people to talk to that have the same qualifications. Sometimes this matters, sometimes it doesn't. For example, career professionals don't normally have additional training to help HDR

Figure 1.4   As a graduate with a research degree (master's or PhD), you have a unique set of skills and considerations. But the basics are still the same. Photo from National Cancer Institute on Unsplash.

graduates specifically. You will likely find just as many helpful points in the other chapters as other science professionals, but we wanted to provide tailored support too (Figure 1.4).

Use of Badges and Icons

Throughout the guidebook, career-related activities are labelled with skills 'badges'. These are icons, which signal that the activities will also help you develop specific transferable skills (also known as employability skills, key skills, 21st century skills and many other names – see Chapter 2: Transferrable Skills and Reflection). Developing these skills is critical because they are needed to thrive and make a valuable contribution to any organisation. These transferable skills also underpin the successful application of your scientific and technical skills. For these reasons, employers value them highly. Transferable skills are powerful because they can be transferred from one job to another, from one employer to another, and across industries and professions.

Examples include written and verbal communication skills, problem-solving skills, critical thinking, time management, organisation, independent learning, numeracy/data analysis, initiative, software/IT skills, adaptability, creativity/innovation and commercial awareness.

The skills badges (see Key to Skills Badges) were created in response to research into science undergraduates' recognition of the transferable skills they were developing at university and those they wanted to develop. It became clear that students were focused on a narrow range of skills and were not conscious of some of the key skills desired by employers.[1] To address this, the badges were displayed on curriculum tasks (labs, workshop activities, assignments/projects) across various science subjects. Research confirmed that the badges helped students recognise a greater breadth of transferable skills development, feel more motivated and satisfied about their tasks, and identify examples they could share when applying or interviewing for jobs.[2] Therefore, we have included skills badges throughout this book in order for you to better understand and recognise where and when you are developing which skills.

Alongside recognising the skills you have developed and opportunities you sought to strengthen and broaden your skills, it's important to reflect on your study and work-related experiences throughout your career. There is significant evidence that reflection enables us to learn more from the situations we experience – analysing and taking meaning from them, identifying what we have done well that we can repeat and what we could do differently in the future to improve (Chapter 2: Transferable Skills and Reflection). In other words, reflection helps us to keep learning and progressing as professionals. Reflection can also increase our ability to articulate our skills and attributes and our related professional development, a key skill for career development.[3] Because of the power of reflection, you will be encouraged to reflect on many situations and aspects of your development as you progress through this guidebook.

Key to Skills Badges

Throughout the book you will find useful hints, summary quotes and activities to check/help with your understanding. Look out for the activities and how long to spend on them. Do return to these at time intervals as indicated. Use the icons to guide you in the skills you will learn by doing the activities.

 Organisation and time management

 Thinking and problem solving

 Use of tools, technology & software

 Teamwork

 Oral communication

 Numeracy

 Independence & initiative

 Self-evaluation

 Creativity

 Commercial awareness

 Adaptability & flexibility

 Written communication

 Graduation

 Developing skills

 Networking

 Developing a CV

 Applying to jobs

 Task will take x minutes

 Review in x months

What is Your Self-perceived Employability?

Everyone has perceptions about themselves. These aren't real measurable points. They are ideas we hold about ourselves. When it comes to careers and self-development, what you think about yourself is really important. This is because people tend to act on what they think or feel, not some measured reality and not other people's opinion. Our first activity in the book is for you to look at your self-perceived employability[4] and file this away for reference later in the book. You might want to use the results during a number of activities, but at the end of the book there will be an activity specifically to think about the difference in self-perceived employability before and after reading the book. We ask you to enter your email only so we can send you the results. If you are uncomfortable doing that, you can just screenshot your answers. Your email will never be used for any sort of advertising, and it is not held by the publisher. If you agree you want to be a candidate in one of our activities, however, we can keep your email and contact you if an opportunity comes up, for example, to talk to us about your employment journey for a new book.

Activity 1.1: What Are Your Self-perceptions About Your Employability?

Activity goal: Explore your self-perceptions around your employability.

Purpose and benefit: In order to understand how you think about your own employability and maybe even why, complete this series of questions (see the QR code), which research has shown is a reliable insight into your perception of your employability. The tool looks at outside influences and internal ones, the influences that you can control yourself. Self-perception, which this tool looks at, is really important because we make decisions based on what we believe and feel about something, not how others see us.

This tool is also for use further on in the book when you want to think about whether your self-perception of your employability is changing and how. There will be reminders every couple of chapters (near the end of the chapter) so you can check in and learn about your learning.

Activity 1.1: What Are Your Self-perceptions About Your Employability? – (*Continued*)

Activity steps:

1. Follow the QR code on the right to the questions. Complete them and get a copy sent to your email for reference, both now and to refer to later in the book to reflect on your progress and better understand how far you have come and how. This is very useful for future learning and future professional development.

2. Review the results you were emailed. Are there any surprises? Note down a few thoughts about either the questions or your responses.

3. Take either a point that you were surprised about or a point that you wish to work on and write 3 bullet points about it. This could be something to follow up on, a point of interest, a random thought, a question to ask someone. There aren't any wrong answers. This process is probably helped by not over-thinking the process or your answers.

4. Put these notes in a file, which you will continue to add to as you work through the book. Make sure you label it clearly with the activity number so that it is easy to find.

Takeaway: Better understand where you are coming from in order to work out where you want to go, and appreciate how far you have come as you work through career development.

Good Luck With Your Journey

This book is a tool to help you navigate the development of your STEM career. As I said at the start of this chapter, different people will use the book in different ways, travel through at different speeds, and revisit different sections at different times. This is totally normal. Most people won't simply read this book from front to back.

All the authors of this book wish you luck with your journey of developing your STEM career, whether that is in data, the lab, fieldwork, as an educator, in research, policy, marketing, sales, consulting or anywhere else. There is no right way to "do" a STEM career. We hope this book can help you develop your understanding of your skills, how to develop and articulate them, as well as helping you understand workplaces and what you want from your career and why. We all wish you good luck with the journey.

References

1. M. A. Hill, T. L. Overton, C. D. Thompson, R. R. Kitson and P. Coppo, Undergraduate recognition of curriculum-related skill development and the skills employers are seeking, *Chem. Educ. Res. Pract.*, 2019, **20**(1), 68–84.
2. M. A. Hill, T. Overton, R. R. Kitson, C. D. Thompson, R. H. Brookes, P. Coppo and L. Bayley, 'They help us realise what we're actually gaining': The impact on undergraduates and teaching staff of displaying transferable skills badges, *Act. Learn. High. Educ.*, 2022, **23**(1), 17–34.
3. M. A. Hill, T. L. Overton and C. D. Thompson, Evaluating the impact of reflecting on curriculum-embedded skill development: the experience of science undergraduates, *Higher Educ. Res. Dev.*, 2020, **39**(4), 672–688.
4. A. Rothwell, I. Herbert and F. Rothwell, Self-perceived employability: Construction and initial validation of a scale for university students, *J. Vocat. Behav.*, 2008, **73**(1), 1–12.

Transferable Skills and Reflection (the Career Building Transferable Skill)

ANGELA ZIEBELL

Deakin University, Australia

What Are Transferrable Skills?

Transferrable skills are general skills that you use in many different settings, which are not specific to your discipline or technical area (see Figure 2.1 for examples of transferrable skills). You might also have heard them called 21st century skills or employability skills. We use the term transferable skills in this book as we feel it best describes the collection of 'people' skills (also referred to as soft skills) that are not discipline-specific, but that you still need to develop throughout your career.

You will use these transferable skills in your personal life too, but we often think about them more when it comes to study and the workplace. Transferable skills enable you to use your discipline skills better and in a wider range of circumstances. For instance, good oral communication skills can help a person make sure they have understood all the technical steps in a new procedure, or planning and organisation can help you manage several technically demanding things together when it would otherwise be impossible.

> "One of the biggest mistakes I see in very clever research students and very clever science students is the fact that they've got the knowledge, but they can't convey it to either a scientific or non-scientific audience."
>
> – Shezmin, chemist and former QC analyst, passionate about STEM outreach and currently working as an application chemist post-PhD.

Like any skills, transferrable skills can also interact and support each other. Someone who has great verbal communication and great planning/organisation skills is likely to be good at helping others plan and organise. Someone with great written skills but poor verbal communication

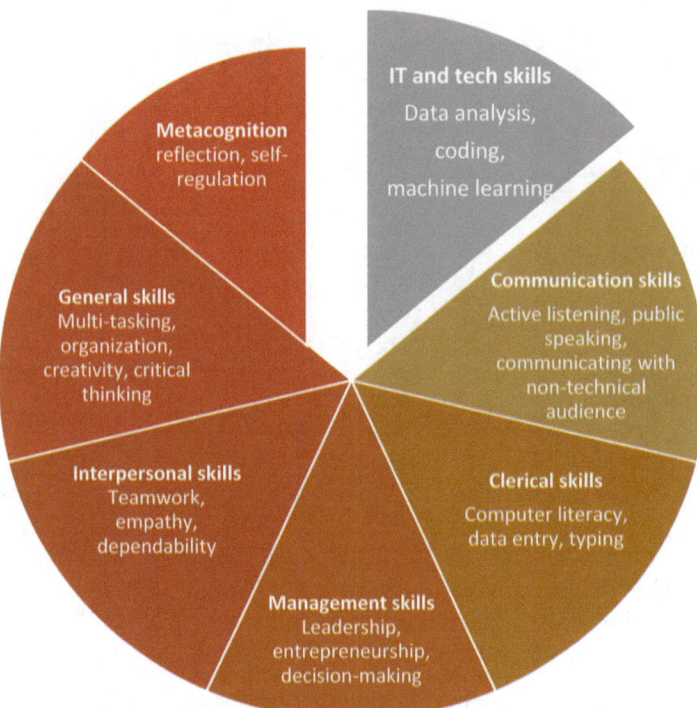

Figure 2.1   Transferable skills cover a range of activities and are often used in combination with one another.

might understand their communication strength and rely mostly on written work. This person could do things like send their thoughts around to people by email ahead of a meeting to avoid being put on the spot during that meeting. Everyone works differently, and variations in transferable skills is one reason why. How many transferrable skills can you notice underway in Figure 2.2?

Why Are Transferable Skills Important?

While discipline skills like coding, safe lab techniques, instrument/machine use, and sample collection techniques are vital parts of a STEM workforce, it is our transferrable skills that we use most of the time. If you take the average experiment or collection of data, you will: plan, use basic software and IT skills, talk/write to others to check details, find additional information from the internet or reference texts (independent learning), manage the resultant samples/data by labelling and saving

Figure 2.2   How many transferrable skills are being demonstrated here, do you think? Which ones? Photo by fauxels from Pexels.

files in a well-labelled and organised manner, and (hopefully) keep your notes up to date and organised while you work through all of this.

All those steps rely on your transferrable skills, not your technical skills. The transferable skills in this list are independent of the exact field the experiment or observations are done within. For example, I could be describing work done in chemistry, biology, hydrology, geology, health science or financial mathematics. You cannot tell from the transferable skills and transferable skills do not work differently in different fields. You can transfer fields and all your transferable skills travel with you.

> *"I think the benefit of being a STEM graduate is you do have a lot of skills that can be applied to so many different areas. So you don't need to be in the lab. You can go into industry, you can go into government, you can be in policy making. Like, there's so many things that a STEM graduate can step into. And I think that's something that is a real benefit to being a STEM graduate, that you're not necessarily pigeonholed into doing a particular thing or being in academia or being in a lab or being in research. You've got so many different things that you can step into."*
>
> – Alex, Bachelor of Arts/Science, master's in bioethics, coordinator of indigenous health education at a university.

So transferable skills are important because they make up a lot of what we do at work and, where they are not the main focus, they help us carry out our technical or discipline-specific role better. As professionals, when a number of key transferable skills are under-developed, it is harder to put those hard-earned technical skills to work. To avoid this and put you in the best position to let your technical skills shine, we need to work on transferrable skills just as much as discipline skills. Sometimes more because they are under-developed. Next, we will look at the skills gaps seen in STEM graduates in a number of countries, and help you work out what skills gaps mean for you. Then we will turn to thinking about your individual skills gaps and how you can work on them. In all these sections, we will help you understand transferable skills in general, but also understand **your** transferable skills so that you are better placed to improve where you want to.

"I'm [in] manufacturing. It is really actually quite technical and using a lot of machinery. In that kind of environment, they're really looking for problem solving and critical thinking. That's why they're looking for engineers and scientists. It's a good job, fantastic to work with people, especially my workplace. We have 400 staff working on the production floor. I feel that they're looking for someone who has people skills and ways of communicating with people."

– Ruth, science graduate working in the retail food industry as a line manager, former business owner.

Skills Gaps in Our Systems

On average, new graduates have transferable skills gaps and studies have found this to be the case in many places all over the world (Figure 2.3).[1-7] It is normal for STEM graduates to finish university with extra technical training they might not use, but without their transferrable skills being at an ideal level.[6] The research shows that employees and recent graduates generally agree on the rough distribution of skills gaps. However, their choice of how employees and employers rank the need to improve individual skills does vary. This variation makes sense because the students and employers will see their skills from different points of view. For example, an experienced employer knows how an employee will improve over time but also knows the skills they will need over time, not just in their first year. An employee is also likely to sense a (bigger) gap when they are keeping up and working at a reasonable standard but using a lot of effort to upskill while working.

Figure 2.3 It's important to be conscious of all sorts of gaps in your learning, so you know where to focus your attention when building new skills. Photo by AXP Photography on Unsplash.

"When I got to my industry, I had to have a very open mind. We mostly just studied theory on campus in Kenya and so what I'm doing now is very different. Everything is a learning opportunity currently. I might go to a different industry one day and learn lots of things from scratch. So when you are leaving campus to work you just need to have that open mind. Wherever you go you are capable of learning. You are capable of catching up to everyone who is already there. And once you catch up, you are capable of doing a lot of good."

– Faith, mechanical engineering graduate in the food industry.

Lastly, graduates know what they learnt at university but, as we discuss later in Articulating Your Transferrable Skills in this chapter, students and new graduates are not always good at identifying and articulating those skills. Therefore, some skills might go undiscovered by the employers. Employers, not surprisingly, tend towards wanting the ideal graduate. You would want the ideal employer if you had a choice too, right? So, it is natural that their list of transferable skills for a future employee is extensive. This might match their hopes more than what they actually understand that they will see when they interview graduates.

Skills needs can also change rapidly as economic or political changes often occur rapidly. For example, changes in import/export agreements rapidly change manufacturing, including advanced manufacturing. Another example is the use of long complicated supply chains that became almost impossible to use during COVID. COVID also gave us an example of when government spending can rapidly change spending impacting STEM jobs. Degree completion timescales are long compared to these processes, so training cannot change at this speed, even in places where degrees offered are mapped skilfully to the country's workplace needs.

Why Do Skills Gaps Exist?

There are several reasons why skills gaps exist, and how much they contribute to skills gaps in different work environments around the world. The most common, in my mind, is that universities have traditionally focused on discipline skills. Traditionally, academics teach their area to ensure students know about that specific technical area. Then another academic teaches another area. You put this all together using a degree plan or requirements, and it is a degree. The academics are fantastic in their area, but their area is teaching and research of their discipline. This system focuses on the assessment of whether the student has understood the content of the area and how to use it (*e.g.*, during labs, presentations, or field trips). Some transferable skills are usually developed in most degrees. For example, group work is common, especially in the lab or in the field. Presenting either in a group or by yourself (online/in-person/*via* video) is also a regular part of most STEM degrees. It is more common that information and communication technology (ICT), communication, commercial awareness, and independent learning are under-represented.

However, with all these skills, there is also a trend for students to under-recognise transferrable skills (we talk about that point quite a bit in this chapter). No matter how the degree is set up, if it's a general science degree, the educators you have do not know what workplace you will end up at. Therefore, tuning the degree to have the right mix of transferrable skills is fundamentally harder. Education systems all over the world are trying to increase the number of transferable skills students learn and how well students recognise their abilities and can transfer them out of the classroom and into the workplace on graduation. Sometimes this means they add a placement or a career skills class. Sometimes it means sending you to work with the central career skills teams (which many STEM students don't know exists).

"In hindsight I should have networked more when I was a student. I took a long time to get my first job and if I had of networked more, I don't think that would have been a problem. I should have taken advantage of the code camps and done side projects to develop my skills, e.g., using free and open-source software."

– Nimaz, software engineer working in education software.

Where Do There Tend to Be Less Skills Gaps and Why?

In degrees like engineering, teaching, and nursing (those that are strongly linked to specific jobs), the link between the workforce and the teaching staff is stronger and many staff are former or current practitioners. In these profession-linked degrees, skills gaps are less likely to occur because the teachers of the subject are often an academic-practitioner. They teach **and** practice their discipline; therefore, they are better placed to understand the exact transferable skills needed, how much they are needed, and how to teach/assess those skills in workplace linked ways. Students are also already headed in a very specific professional direction, so it is easier to teach to a much less diverse cohort of students.

Another feature of the university system that can bring educators closer together with employers is how much the funding environment encourages researchers to do research with businesses, and how many businesses there are that could use their skills. This varies hugely from country to country. To learn more about your situation, I encourage you to find out what the situation is in your area or (if you are studying away from home) it can be good to investigate what the situation is at home. You could do this just by general internet searches, talking to graduate students or researchers, or talking to a graduate already working in that country. Do not know any? To find out ways of developing relationships with people, Chapter 4: Networking will help a lot.

Recent Changes to Include More Transferable Skills Teaching

The inclusion of transferable skills is changing over time though, with the recognition of the importance of transferable skills in the university curriculum. These days, some degrees offer significant chances for students to develop their transferable skills. For example, some degrees offer frequent, well-structured group projects, working on industry-sourced materials or data, attending internships as part of the degree, completing research projects outside of the university, entrepreneurial challenges, and fieldwork modelled around real scenarios. Other degrees might not have offered any of those.

Another related opportunity is taking an interdisciplinary subject if that's possible in your degree. These classes get you working with people from different degrees and are generally great for building transferable skills. In part, this is because you will learn a lot about working with different people, just like in a job. If you are looking to change degrees, start a new degree or offer advice to a potential student, be sure to consider the ways in which the candidate universities help students develop their transferable skills *via* these options.

Some graduates find it is only when thinking about how and what they were taught after they start work that they realise they were being taught some skills. They were actually being taught the skills, but they did not realise. This highlights the importance of thinking about what you learn(t) and where that learning sits in your wider education. We will cover ways to do this later in the chapter when we talk about reflection.

In some areas there is a long history of hiring people based on marks and university/department name. In most countries and industries marks are only just part of the picture and in the last couple of decades are slowly becoming less important. But it should be recognised that this varies greatly, and it is a valuable point to investigate for your discipline and location. More commonly, marks are weighed up against your transferable skills, your work experience (professionally linked and not professionally linked), how interested and prepared you are for the application and what you want out of job. These details will be talked about more in Chapter 6: Resumes and Cover Letters and Chapter 7: Addressing Key Selection Criteria and Interviews, when we look at the application process and interviewing.

Your Transferable Skills

Articulating Your Transferrable Skills

Some of you will have great transferrable skills. What we find though, is that many students and graduates do not yet understand either what their transferable skills are, or how to talk about their transferrable skills when looking for positions. In such a situation, those students and graduates are not submitting the best applications that they could, and they are not giving the best interviews that they could. Therefore, they do not have the best chance of getting the jobs that they could if they better understood the skills they have and how to talk about them. This can impact applications for a whole range of types of roles, *e.g.*, internship, volunteer, paid, professional or non-discipline linked. I certainly did not know how to talk about my transferable skills when I was a student or a new graduate. Applying for jobs is not an intuitive process, so reading and taking time to understand the process is very helpful. First, we will talk about understanding, articulating, and building your transferrable skills. Then, we will look at how you can maximise what you have learnt about transferable skills in the job applications and interview process (Chapters 6 and 7).

"That's [sic] one of the things I find is really one of my advantages is that I've worked in a broad range of industries and places that I'm able to sort of adapt depending on where it is. And there's a broad range of soft skills that regardless of wherever you work, when they say teamwork, critical thinking, ability to problem solve, ability to communicate etc., regardless of where you are, it will always be important."

– Nigel, science physiology major, worked in a bank while a student.

All professionals gain from learning about how to articulate their transferrable skills when they apply for jobs. The first thing to learn is that, yes you do have to articulate your transferable skills. In STEM it is common for professionals to focus mostly on discipline skills and/or to focus on job descriptions or outputs. None of those describe what you are like as a professional. When it comes to your job applications, if something is not listed and nicely explained then it will be assumed to not exist. So learning to write about, and talk about, your transferable skills will put your application higher up the list each time. This will get you on more shortlists and get you into more interviews (Chapter 7 takes you through interview preparation), leading to the best chance of getting a job you want.

Talking About Your Skills Can Be Uncomfortable

Some readers are already uncomfortable at this point because the thought of putting their skills forward is not something they are comfortable doing. Keep in mind that we are asking you to say what you feel your strengths are. We are not asking what you are the best at. You do not have to be the best at something for it to be very valuable. Employers also very much understand that having someone who is great at a range of skills is almost always better than having someone who is amazing at one thing and does not want to do other things or is not good at other things.

Being uncomfortable with talking about your personal strengths seems to be very common in STEM fields. Students and graduates seem very comfortable to write that they can use a piece of equipment or that they were taught X, Y and Z technical methods. But, on average STEM people deal with facts and data so the idea that you can just **claim** you are good at some immeasurable skill can seem wrong or counterintuitive. When you talk about your transferable skills, however, you are not just going to be saying "I'm amazing at this". You need to be able to talk about your skills, how you developed them, and why. It is about explaining your skills, not bragging or "selling" yourself.

Hint: If you cannot think of a transferable skill to write about this might be related to not being comfortable with the idea of saying you are good at something. Try thinking about what you are less bad at. These will be the things others think you are good at. Not having the confidence to say what you are good at is pretty common and happens for a range of reasons. There are several places in the chapter where we will give advice to people in this position.

For example, I had quite good data entry and analysis skills as a student after my second year because of a specific role I chose to play in a group project. I was in a group that had an ecology report to write up from a field trip. I decided to commit to doing all the data entry and analysis on Excel as I my Excel skills were not very good, but I knew it was important to improve. Making myself do lots of data entry and analysis was a good way to learn. I was a little scared whether I could learn what I needed to, but I reassured myself that, seeing as others had learnt, I could do it too. After this project I made sure to keep using Excel so that I did not forget my skills.

"I think when I started working in my human resources job, I started to real-ise how often we use things like Excel and analytics and power BI, SQL. And I was like, oh, these things are widely used and they're becoming, you know, it's becoming more prevalent in the workforce to use data."

– Marina, psychology graduate working for a consulting company.

In my first interview after graduation, I told them the short story of how I became relatively skilled at Excel (I never claimed to be an Excel genius, make sure you do not over claim). The interviewer could see exactly the experience I had had, and it was evident that having that sort of experience would give me good Excel skills. Could I have been lying? Sure, but really, who would make up that story? And if I was lying, then the discussion we had about the experience would have felt wrong in some way, and the interviewer would have been wary of trusting my claim. It is the conversation you can have about how you developed the skill, the things you can now do, and what you have learnt that come together to show that you have a skill. While this is short of actually demonstrating a skill to a person, it is so much more than just saying that you have a skill.

So let us think about articulating a transferable skill you have in Activity 2.1, then later we will work through some activities that help you more fully expand your transferable skills list.

Activity 2.1: Explain (Articulate) a Transferable Skill and How You Developed It

Activity goal: Learn how to articulate a transferable skill you have and how you acquired that skill.

Purpose and benefit: Learn the basics of how to write about one of your transferable skills in order to prepare for working on your job applications documents (build on it in Chapters 6 and 7). This activity will also help you better understand your transferable skills so that you can work out what skills you want to develop further and why.

Activity steps:

1. Pick a skill from the list in Figure 2.1 that you could write about, or if you have another skill in mind, you can use that too (quickly use an internet search to check if it is considered a transferable skill first though).

2. Write that skill as a heading and below answer these three questions.
   - How did you develop this skill (there could be multiple points)?
   - What is it that makes you think you are good at this transferrable skill? Make sure at least one of these points is one of your observations, not just what others have observed about you.
   - If someone asked you about this transferable skill in an interview and you had 30–60 seconds to respond, what would you say? Why?

3. Once you have worked through this, find someone to talk to about your transferable skills. Talking to some else will also be part of later activities in this chapter so you might want to identify a person to talk to and then work through a few more activities before you meet.

4. Summarise what the other person said and identify:
   - what you learnt from the conversation,
   - why you learnt it and,
   - what you are going to do with that information and why.

5. If you are interested you can go through this activity with 2–3 transferable skills. This will be particularly useful if you are having trouble working out how to write about your transferable skills.

6. Either because it feels more natural or because you want to improve your oral communication, you could also turn this activity into a discussion if you have someone you can talk to about this point.

7. Return to this activity as you gain skills to:
   - redo the activity with a higher understanding of transferable skills to improve your articulation and understanding of your transferable skills and/or,

Continued

Activity 2.1: Explain (Articulate) a Transferable Skill and How You Developed It – (*Continued*)

- see how your thoughts have changed, which will help you understand what you have learnt.

Takeaway: Being able to write about your abilities is an important part of many aspects of being employed. As you practice, you will also get to understand your transferable skills better, improving your articulation (your ability to explain and talk about your skill). While different workplaces function in different ways (*e.g.*, in some workplaces once you have a job it is more common for senior staff to identify who is good at what, others are more or less competitive) there are many times when articulating your transferable skills well is important. This can include applying for new jobs (see Chapters 6 and 7), going for a promotion or a new role in the same company, networking in general (see Chapter 4), or trying to make a case with a particular person in the company about your abilities.

Later in Chapter 6, we will talk about exactly how to include your transferable skills in your job application and how to bring them into conversation during an interview (Chapter 7). For now, we will run through a few exercises to get you thinking further about your strengths and weaknesses, so you can plan what your next development steps are, given the area you are interested in working on.

You Can Develop Transferable Skills Anywhere

As I said in the introduction to this chapter, transferable skills are used in all parts of our lives. That means they are developed in all parts of our lives. They are transferable enough so that the customer service skills you learn working at a restaurant or a shop can support your entry into a STEM job that requires customer service. For example, consulting or technical sales positions both need people that understand how to talk to, and deal with, customers. Commercial awareness can similarly be learnt working in a kitchen, where serving sizes, and whether all the food sells before it turns bad, is central to whether the business survives.

You can also develop transferable skills in roles where you are not paid. You might help to run a student club, care for a grandparent (Figure 2.4), help with the family business or volunteer regularly. All these roles are very important, and you can develop similar, if not the same, transferable skills as if you were getting paid.

Figure 2.4   Important transferable skills can be developed by helping family or volunteering in your community. Photo from Pexels.

Working for a family business will almost always help you with organisation, commercial awareness, and customer service. Working in a volunteer role, you can really develop any skill that you would develop in a paid role. If there are permanent staff in a similar but more senior role, you are likely developing all the skills they need for their role. The fact that you are donating your time to help does not change how you develop your skills.

Helping care for a loved one will teach patience and empathy. But these are not transferable skills, no matter how important they are. Look at things like: are you in charge of their schedule? Do you have to help with medication and food? Do you have to understand and keep track of diet or medical information? From just this list of tasks, I would say you were (at the very least) developing your communication, organisational, and customer service skills, and probably independent learning.

Experiences Do Not Have to Have Fancy Titles to Be Meaningful

In many cultures, there is a lot of prestige attached to certain experiences, *e.g.*, internships at well-known companies. But these experiences, by design, are usually only for a few. They are also often decided by who you know, or by your absolute mark. They do not always take into consideration other important characteristics that might make you a great professional. Additionally, some people might find they are overwhelmed by the thought of a "big" experience that could mean things like moving inter-state. Thus, it is common for these roles to go to the more confident students and graduates. However, confidence alone is not a great predictor of what a great professional you will be. Finally, it is very common for these experiences to be unpaid or minimally paid so that they are only available to those that can afford to give up time at a regular job or potentially resign. So if you find a fabulous, fancy-looking opportunity, go ahead and take it, do not let me detract from the wonderful experiences these can be. But if you cannot take the position, there is an almost endless supply of other things you can do to expand your transferable skills meaningfully.

> *"I think employers like really well-rounded people. So I suggest students do extracurricular activities to improve networking and communication so you will be more attractive to future employers. When I was a tutor I learnt how to break things down and explain things, as well as approach things differently depending on who I was working with. These are really important transferrable skills."*
>
> – Alex, research scientist at a microbial protein production firm, born in the US, has lived, studied, and worked around the world.

But I Do Not Have Any Work Experience!

If you have a background with little or no work experience and you are planning on applying for professional roles in the next year or so, now is a good time to look for a role. Chapter 5: Portfolio Building is a good place for you to start reading to work out your options and work out how to proceed while still a student. Chapter 4: Networking is your next stop as you start to work out how to navigate meeting people so you can learn more about roles, industries and important issues that impact your discipline or industry at the moment. While it will take quite a bit of work, between these two chapters and a visit to your careers office or student support, you will be in a good position to fill your work experience gap. This is an excellent project for a holiday break, but any time is useful. You can make progress by simply putting aside an hour each weekend to read, reflect, plan, and apply for a position. It is also important not to over-think these things. Information might come up during a class or a casual conversation, you might learn something career-related from reading the newspaper. Once you are thinking in this space you can learn from a huge array of different sources. And remember, there is no one way forward. Everyone's path is different.

What Are Your Transferable Skills Strengths and Weaknesses?

It can be valuable to map out what you think your transferable skills are. There is no right or wrong way to do this. To help you, we have constructed a quick activity (Activity 2.2). The activities after Activity 2.2 build on this first one. Do not be surprised if your answers in this activity change regularly as you progress through your degree and into the early years of your career. That period of time is very changeable, and it is when we usually learn most quickly about the workplace and about how we fit into the workplace.

Activity 2.2: Mapping Your Strengths and Weakness

Activity goal: Map your transferable skills strengths and weaknesses using this activity.

Purpose and benefit: This task will help you understand what transferrable skills you perceive as your strengths and weaknesses. The activity only takes a few moments to do and you can track progress over time by keeping a record.

Activity steps:

1. For each of the areas of transferrable skills in Figure 2.1, pick at least one. To what extent do you agree or disagree that your experiences up to this point prepared you for this future job? Apart from some advanced IT and project management skills, the skills on the list in Figure 2.1 are transferable skills. These are meaningful for most, if not all roles, making them especially important to be conscious of developing and being able to articulate.

2. Once you have written down your thoughts on those skills (at least one from each wedge of the pie), think about what the results mean to you. Are there some surprises? Are there things you want to work on? Points you are proud of, or relieved to realise you are comfortable with?

3. At this point we are not writing anything like an improvement plan. Your aim is to understand your strengths and weakness. Label which ones you think are strengths and weaknesses and have a think about what order they might come in from weakest to strongest. Does that reveal anything? There could be a pattern, for example, that you like to spend time improving on things you are naturally good at, so you have gotten really good at those. Or maybe you feel challenged trying to claim you are decent at some of these.

Continued

Activity 2.2: Mapping Your Strengths and Weakness – (*Continued*)

4. Like most of these activities, if you get stuck or want to get additional ideas, we encourage you to talk to someone else you trust. They do not have to have a lot of STEM knowledge to help you think through something like this. It might even be that just talking out loud to someone helps you with your thinking. They can also be an independent measure. They can tell you what they think "good", "excellent" or "needs work" might look like in the workplace.

Takeaway: As you improve your transferable skills and improve your understanding of your transferable skills, you can revisit this activity to monitor your progress. A good timeline would be every six months while you are actively working on developing your transferable skills, including your understanding of them.

Keep in mind that a need to further develop transferable skills is not a bad thing. Unlike degree-specific discipline or technical skills, transferable skills can be developed in a lot of environments. Essentially anywhere. You can develop amazing transferrable skills outside of your studies by volunteering, working in a family business, getting a part-time job, doing an internship, or just finding everyday opportunities to practice a skill. For example, someone wanting to improve ICT skills might challenge themselves to be the person to gather, input and format numbers next time it is needed. Or someone who is uncomfortable with teamwork might take a part-time job in a team-based role deliberately to build teamwork skills. If you are a student reading this, then you likely have time to fill a number of skills gaps before you graduate. If you are already a graduate, then you can fill skills in a similar manner. This might also include taking roles at work that extend you in an area you want to develop. Or you might get the chance to do a course while you work to expand a skill. This may or may not be a course offered by work. There are plenty of free online learning platforms (*e.g.*, MOOCs a.k.a. Massive Open Online Courses, LinkedIn learning, Kahn academy, Duolingo). Studying while you use the skill in your role is a great way to learn quickly as you are using what you learn regularly, and you understand why you are learning, which tends to be a very engaging way to learn.

"I think in terms of my degree, there was a lot of transferable skills in terms of you learn routine, you learn organization, you learn time management."

– Ojasvi, chemistry/physiology graduate. Formerly arts/science student, clinical trial clinician.

Building these skills can at times be a bit uncomfortable, but it can also be really rewarding. For example, it is very common to work out that you are not as bad at the skill as you thought you were. Some people discover a new side to the skill they thought they were weak in. For example, you could realise that it was more confidence that you lacked rather than skill, then you can try to work through the confidence issues. Or you could get a better understanding of how to work with a weakness. I learnt in my first job that, while I really liked being organised, it only went so far. I was not a good match for the very controlled environment of quality testing, so I have made sure not to work in very regulated areas again. This was a positive for me, but it is also a positive for anyone who might have employed me in that area. They would have hired a poorly suited employee that was unhappy and likely prone to mistakes. I should add, it is also a positive for anyone who would have worked with me if I had stayed in that environment. I would not have been fun to work with.

Now, for Activity 2.3, I want you to think further by looking at some job advertisements and finding where your transferable skills are listed. Keep in mind that it does not matter if the positions are a bit more senior than you would look for, or in a city you do not want to move to, or even if there are discipline differences (*e.g.*, chemical technician *vs.* biology technician). This does not matter for this activity because transferrable skills are used everywhere.

Activity 2.3: Identify Transferable Skills in a Role That You Might Want

Activity goal: To be able identify the ways transferable skills are included in job advertisements and identify a skill for you to develop.

Purpose and benefit: This task will help you understand where given transferrable skills are important, and help you to start thinking about how this interacts with your career plans/thoughts and your transferrable skills strengths or weaknesses.

Continued

Activity 2.3: Identify Transferable Skills in a Role That You Might Want – (*Continued*)

Activity steps:

1. Keeping in mind the transferrable skills terms from Figure 2.1, go to your favourite jobs site (if you do not have one in mind search the internet to find out what are the most popular job search websites in your region and use one of them).

2. Put in a number of general job search terms so that you get lots of hits (the precise job is not very important here). Some ideas might include just searching for scientist, engineer, laboratory technician, graduate scientist, *etc.*

3. Find five jobs that look like they are roughly in your geographical area or an area you plan to work in.

4. Read those five job descriptions and highlight any transferrable skills that you find.

5. Take 10–15 minutes to write a ~100 word description of your abilities in 3 different transferable skills you found in those position descriptions (very roughly a total of ~300 words). Leave a few lines of space under each for notes.

6. Now answer the following questions to help you further think about the relationship between the skills and the roles.
   - Did the roles have transferrable skills in common? If so what and why do you conclude that is?
   - Do you think you fit the description in the advertisement in terms of the level and quality of that transferable skill that you currently have? If yes, great.
   - If no to the above question, what ideas do you have for improving that skill?
   - Is there an obvious mismatch between these jobs and your skill set? Are there related roles that have other skill sets that might be better suited? You can find this by looking through a wider array of roles and searching for those skills. Or you can try an internet search with your discipline (*e.g.,* programming) and then the skill(s) to see what comes up.

With regards to transferable skills, what surprised you about what was in the advertisements and why? Or why not if you weren't surprised?

Hint: If you had trouble finding roles that help with this activity, you can visit your university career support area, including after you have graduated in many cases. Ask your favourite academic or the university student support group if you can't find the careers help staff.

Takeaway: We will discuss exactly how to present and express your transferable skills in your resume, cover letter and key selection criteria response, as well as when you are interviewed, in Chapters 6 and 7. For now, the aim was to start you thinking about how transferable skills are used to select people into roles and about how and why your skills might be suitable for different roles.

How Do Others Grow Their Transferable Skills?

Now that you understand your transferable skills a bit better, it will be useful to get input from others to better understand transferable skills. In many of the activities in this chapter you need(ed) to find someone else to talk to in order to get further input. Here you will also need someone, or a couple of people, to talk to. You might even merge these together into the one conversation.

So how can we grow our transferrable skills from the figurative caterpillar in a cocoon to the beautiful butterfly (Figure 2.5)? If you are a student, you might have a study group or a group of like-minded students you can get together with and talk to. While students are not the most experienced people, other students will be able to look at your skills and abilities and comment on what they think your strengths and weaknesses are. This will especially be the case if you have been on student projects, multiple laboratory classes or study groups together. These experiences are ideal environments to observe the use of transferable skills in others, whether or not you realised it at the time. Importantly with other students, if you are going to talk to them about transferable skills, they will have to understand the concepts involved. This can be as simple as lending them this book. If you are lucky enough to be a young professional with a group of other young professionals you could have this conversation with, then this paragraph also applies to you. Otherwise, you might work with a more varied network of people, including those that you did your degree with, colleagues, friends and family. Students can, of course, work with that same network too, but you will find in

Figure 2.5 Grow your transferable skills. Photo by Suzanne D. Williams on Unsplash.

the last year or so of study there are many fellow students interested in unlocking any progress to apply for employment successfully.

People from the wider community have a much more varied set of experiences, which can be a great thing. But it is also important to think about a person's experience before working out how to receive or use their advice. Not all advice is equal and personal goals and approaches to careers can vary radically. Before you ask to talk to someone, think about the type of jobs they have had. Things you might think about are whether they have successfully changed roles, which means they are more likely to be skilled at understanding your jobs market. However, note my use of "successful". Someone who has just changed roles a lot might be moving for reasons you do not want to emulate. Also, their field might be of importance. If they are in a completely different field (*e.g.*, artistic design *vs.* chemical engineering) then they might have less you can learn from. But do not assume this is the case, just be conscious that there are differences between industries and fields. To understand what is transferable between their experiences and your experiences, ask questions to better understand the roles they filled and the experiences they had. The chances are you will also learn a lot in the process anyway. Another consideration is if they have had to supervise or hire young staff (whether in a STEM field or not). Those that have experience with hiring or with supervision of new, young staff, will likely be very familiar with the concepts in this chapter, although they may well use different language.

> *"A lot of people think there will be a job at the end of the line that looks like my major. And I'm somebody who nobody in my family before had graduated from University. There was no perspective on University. It was only towards the end of my degree, at the start of this career journey, that I've realised how many people do a degree in an area, but then get a job that's not even related to your major, because ultimately the skills from a science degree in particular are transferable to anywhere. Like I said before, it helps you learn. It helps you think critically to question to be a really good thinker and participant in an organization."*
>
> – Dylan, chemistry and physiology, went into finance after an honours project about financial decision making.

Once you have identified a person or some people to talk to, structure some questions to help identify what they see as your strengths and things to improve, and why. Not sure where to start? Take a look at Activity 2.4. Do not assume people know what transferable skills are when you refer to them as transferrable skills.

Activity 2.4: Asking Others About Your Transferable Skills. Consider With Activities 2.2, 2.3, 2.5 and 2.6

Activity goal: Plan and carry out a conversation with someone to help explore your transferable skills, including your strengths and where you can improve.

Benefit and purpose: Further understand your transferable skills but also start to understand how others see them too. We are often our harshest critics, but even when we are not, other people's impressions can give us new insights into our own skills.

Activity steps:

1. Prepare an explanation about why you would like to talk and what you would like to talk about when you are organising to meet/talk. This will allow the person/people to think about the topic beforehand, but most importantly, it will make sure the topic of the conversation and why you want to talk about it, is clear.

2. Think about how you want to talk to the person about transferable skills. Is this someone you want to do some of the other activities with, for example? Is this someone who will need a lot of background information, but they will still understand that individually transferable skills are important? How does this person like to work? Are they quick and efficient or will they like lots of time to talk around the topic and share related thoughts with you. Plan accordingly.

3. What are you going to ask them? Are you going to tell them what you think your strengths are before you start the conversation? Or where you want to improve? Or are you going to get their thoughts first so as not to bias them. How long do you want to talk? *etc.*

4. Think about what you most need help with and write down three main points. Do you want help identifying what strengths and weaknesses you have, or how to improve? Or potentially how to work out where to improved based on a general or specific career direction you have in mind? Talk to the person/people you have selected about these three points to see if they can help you gain any insights.

5. What are the top three things you take from the whole conversation? Why?

6. Pick one of the points from step 4 and work out a plan to develop that transferable skill. You might want to research additional ways to improve the skill, or you might have enough information from the work you have done in this book already. Make a 1, 3, 6 and 12-month plan (or set your own schedule based on relevant events like exams, finishing your degree, big project commitments at work, *etc.*).

Continued

Activity 2.4: Asking Others About Your Transferable Skills. Consider With Activities 2.2, 2.3, 2.5 and 2.6
– (Continued)

7. What are two things you would do differently if you were preparing for a similar conversation with a different person? Why? This step will help you if you have a discussion like this again, but it will also help you prepare for being able to have similar (although less structured) discussions with people at networking events. Networking events are great places to ask people how they developed skills and get ideas. It's a great conversation starter or extender (see Chapter 4 for more details).

Takeaway: This is a type of discussion you can have many times with different people in your personal, social, or work worlds. Often, we do not realise that we can gain information in this manner. As students and junior employees, it can be easy to worry that people have better things to do than answer your questions. However, most people want to help, and you will likely find many relationships growing stronger from such conversations.

What Transferable Skills Do Others Value and Why?

You have thought a lot now about what your transferable skills are and even how you might improve them and why. But have you thought about what others value in terms of transferable skills? And why that might be? People's personality, their own skills, and the job they have influence what transferable skills they prize most highly (although most informed people will appreciate all skills). A designer cannot design without creativity and a project manager will place a high value on organisation and self-management. You can understand people a little based on what they need most for their role. You can also start to get a sense of what some employers will value highly based on their skills gaps within their company. Let us think about these points in a little more detail.

"I think the main ones that employers look for in a grad, is [sic] the ability to learn and have a good attitude and be able to take on new things, and communication as well. Also managing, coming from a consulting background, which is probably kind of teamwork, and things like that. And maybe also being a bit more strategic about how you think about things in terms of your career and how you fit in in the organizational structure, and stuff like that."

– Grace, former actuarial consultant and data scientist/analyst.

Now that you have an understanding of what at least one or two people think about transferable skills, it is a good time to ask them how they have developed their own transferable skills. While you have a captured audience for Activity 2.5, talk to them about how they have developed their own transferable skills. If you find you need prompts, try Activity 2.6.

Activity 2.5: What Transferable Skills Do Others Value?

Activity goal: Find out more about what transferable skills are valued and why by those with a work background.

Purpose and benefit: You will better understand how valuable transferrable skills are, why they are valued and how varied people's opinions are and why.

Activity steps:

1. Find 2–3 people to talk to (independently or together) with at least the equivalent of 1 year's full-time employment per person. If there is someone in the group that has experience in the area(s) you would/might like to work in, then that is a benefit but is not very important.

2. Explain the term "transferrable skills" to them and ask if they have heard this or a similar term (they may not have). What have they heard? Have they used the concept before? How was it used? What do they think about transferable skills as a concept?

3. Ask them what they see as the most important transferable skills in their role or past roles. Why and how did they come to that conclusion?

4. Do they, or did they, see anything changing in their industry with regards to transferable skills? What, how and why?

5. If they could pick one, what transferrable skill would they be better at and why?

6. Feel free to add in as many additional questions that you have as you are talking to them. This is just a set of questions to start the conversation.

7. Given the above discussions, what further insights do you have around transferable skills? Why? What influenced those thoughts? Are you going to do anything with those thoughts? What?

Takeaway: An understanding of how others see transferable skills is helpful when you are trying to navigate and develop your own. See what others are thinking and why. As you learn more about any concept, stopping to think what it means to you, why, and what you will do with that new information can help a lot with learning.

Activity 2.6: How Have Others Developed Their Transferable Skills?

Activity goal: Discover some techniques others have used to build their transferable skills.

Purpose and benefit: By having discussions with others about how they develop their own transferable skills, you will better understand how transferrable skills are used in a range of environments.

Activity steps:

1. If you have not talked to this person about transferable skills before, check if they know what you mean by transferable skills before you start. Most, if not all, professionals will recognise the idea of transferable skills, but they might not recognise the phrase "transferable skills".

2. Ask the person or people you are talking to what transferable skills they have had to work at the most to develop and why that was.

3. Once you have a good understanding of the skills they developed, ask them how they developed them and what advice they would give themselves now if they could talk to their younger self. Why would they give this advice?

4. Feel free to add in as many additional questions that you have as you are talking to them. This is just a set of questions to start the conversation.

5. Share with them what your thoughts are about your transferable skills strengths and weaknesses, and any challenges you see for yourself given your situation (this could be career goals, personal circumstances, industry interest, unusual skills sets or combinations, *etc.*). Being able to talk in moderate detail about the roles you want to go for will help here but is not required. Ask the person you have found to talk to if they have any thoughts about how you could improve your transferable skills given the situation you are in.

6. After the conversation, write out two insights you gained from the conversation and note down why they might be important to you.

7. Given the whole conversation, what are two things that you will act on, how you will act on them, and why? Set a reminder in a few weeks, and a few months, to check back in so you don't forget you had a plan.

Takeaway: Talking to others about their transferable skills and how they have developed them can help us understand both our own transferable skills and the wider world of transferable skills around us. Every employee or business owner works surrounded by other people, and the interactions between the different skill sets influence a large part of how we work. Being able to identify people's strengths and weaknesses is a great interpersonal skill. It can be as simple as knowing who to remind to prepare for tomorrow's meeting or as complicated as trying to work out how a group of five very different people work together.

What Transferable Skills Are Most Needed for the Careers You Are Interested in?

A big part of everyone's career journey is working out where they fit. What transferrable skills you excel at will be part of that picture. As you think about what you might want to do, look at jobs and look at the transferrable skills they are highlighting. What do those types of roles tend to highlight? Which transferrable skills are higher up the list or have a lot of related skills (*e.g.,* verbal communication and written communication, or critical thinking and analytical reasoning). Poorly written advertisements can just look like a never-ending list, but that is why I suggest you look at many to get an idea of what advertisements look like.

You will need to learn more about transferrable skills to understand your own better, to understand what roles are most likely to require which transferrable skills, and how to work out which transferrable skills are needed by a specific position once you have one in mind.

Understanding Your Transferable Skills

It can be very important as a professional to understand your transferrable skills, work to develop them, and understand other people's transferrable skills. One important thing to remember is that, while we all have strengths and weaknesses, we can also all work on weaker points to improve them. They should not be thought of as fixed, although we all have some that can be harder to improve than others. However, you cannot work on these weaker points until you understand transferable skills better. This includes understanding what skills you want/need to improve on, how, and why.

You may have noticed that during the activities, I often asked you an initial question and then follow up with more questions. For example, how or why or when? These prompts are to get you thinking more deeply about the original question. These are reflective prompts and although you might not have realised it, many of these activities were set out to encourage reflection. What is reflection? Well, in the last half of this chapter, we will teach you to work through the reflection process yourself.

Reflection can help you work out what you want and why or narrow down broad choices into much more focused ones. Reflection can help you learn from mistakes, working out how to make sure you do not repeat them or improve a strategy that worked but could be even better. In the next section, you will learn about reflection as a bit of a career development superpower. If you want, as a warmup, think about what you can do today that you couldn't do a year ago (Figure 2.6). This should help remind you that we are always learning.

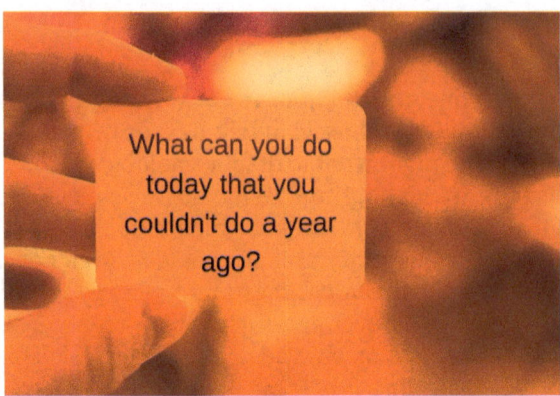

Figure 2.6 Take a moment to think about how you have already been building your skills year on year. Photo by Miquel Parera on Unsplash.

Reflection: A Transferable Skill for Developing Your Career

To think about the value of reflection, let us first think about what happens when people have poor reflection skills. We all likely know someone that keeps making the same mistake. The friend who always eats too much or can never arrive on time. The cousin who is surprised when they don't know what's happening but aren't checking their messages. These people are not learning from their mistakes or using that learning to change what they are doing in order to get a more positive outcome. Habitual change is hard and so some things (like not overeating delicious cake) are genuinely hard while still being obvious. Other points to reflect on are a little more hidden and require thought to work through. However, after we understand the issue fully and work on it a little, we see true progress. For example, maybe that cousin realises that six email addresses and five social media accounts are too much, and they need to retire a few, or more than a few. After a change in the number of accounts, they might now be able to keep up with what is going on. They could originally have thought they were disorganised, but by talking to friends they might work out that too many accounts are the issue and they never realised.

It should be said here that some things will take a lot of work and others will be surprisingly easy. Which points will be easy to work on and which points will be hard is impossible to generalise because everyone is different and we want to get better at things (organisation, communication, long-term planning, fitness, *etc.*) for completely different reasons. So, as you go through learning about reflecting, don't be too hard on yourself.

Who Would You Hire?

Picture yourself hiring someone into a pretend company. It can be whatever you want it to be (check the prompts in Figure 2.7 if you want some variables to keep in mind). They are junior and new to that sort of role, but that is fine because every now and then your company needs entry level people for the more straightforward roles. These people grow into the role, learning new skills and how the company works, as well as the company culture. If this person is a good fit, then they learn a lot over the years and often stay with your company.

Picture two different people and decide who to hire, given all their technical skills are the same. One person demonstrates an ability to think about mistakes, why they happened and how to avoid them next time, *i.e.*, reflection. This person talks to others to work through what went wrong and why, as well as how they can avoid it next time. You know this because during the hiring process, they write about their problem solving on their resume, and then you talk about a similar example in the interview as an example of learning on the job. Would you hire this person, or the person who does not demonstrate an ability to solve problems and improve their actions? It is not a hard decision, is it? You are going to pick the person who has shown they can grow and become better at

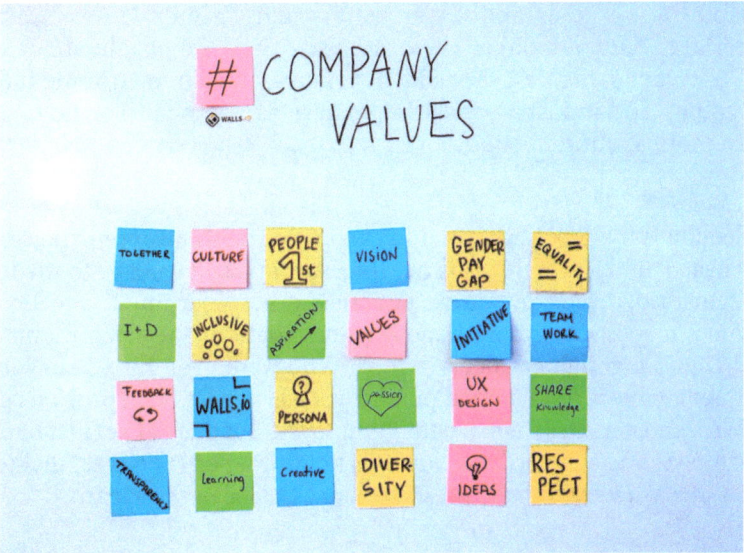

Figure 2.7 What values would you look for in a company? Which ones would be vital, and which values would be optional extras? Photo by Walls.io on Unsplash.

their job, avoiding repetition of unnecessary mistakes. What if the second person has had the same experience but is just terrible at including their good work on their application? Nothing is included. They do not get hired because, essentially, they kept their abilities a secret (although it was unlikely to be deliberate).

What that first person is displaying is the ability to reflect on mistakes and improve, as well as an understanding of their skills. Plus, they understand the need to let their future employer know of their abilities in the hiring process. Is any one part of this process hard? Not necessarily, although it might seem like it at first if you are unfamiliar with applying for jobs.

Most or many STEM students/graduates are new to expressing their transferable skills and reflecting on what they want and how to improve their skills. I certainly didn't know what to say in my first interviews. The way forward is just to learn one thing at a time and soon some things will start to feel familiar. Once thinking about and building on your transferable skills through reflection is not new to you, you will also be more comfortable talking about your transferable skills or sharing your thoughts on them. Being familiar with your skills will help you prepare to write applications and to give job interviews. In later chapters we talk more about how to talk (Chapter 7: Interviewing or Chapter 4: Networking) and write (Chapter 6: Resumes and Cover Letters and Chapter 7: Key Selection Criteria) about your skills. The exact way you write or talk about yourself, and your skills, can vary a lot depending on your environment (both geography and industry) and who you are as a person. For now, let us learn more about reflection.

What Is Reflection, and Why Should I Develop It?

Reflection is the process of thinking about and analysing our experiences. It is based on the fact that it is not the experience that is key to the learning.[8] Instead, "reflection turns experience into learning... and enables learners to gain maximum benefit from situations they find themselves in".[8] This is highlighted nicely by the "Who would you hire?" question in the previous section. The first person has learnt a lot from previous experiences, and it makes them a better employee. But this did not just happen. Learning is an active process, and it is not the experience that "makes" us learn. It is what we do with the experience that "helps" us learn.

> *"Reflection turns experience into learning... and enables learners to gain maximum benefit from situations they find themselves in."*[8]

There are many structured ways to reflect, which can help when we are starting to develop our reflective skills. Here, we will explain two structures that we think are the easiest to pick up for most people, while still being usable in all situations. Some of you will find the process of reflection quite familiar and realise that you have been doing versions of this process already. By learning about exactly how reflection works and the different features, you can start to reflect more purposefully or more fully, therefore getting more out of the process. You might also be able to help others with their own reflection eventually, but no rush.

Reflection helps us learn from mistakes, build on success, re-evaluate our knowledge, connect experiences from different parts of our life together, clarify and enhance what we have learnt, remember our learning better, and make changes and improvements going forward.[8] As I said, reflection can be a kind of career development superpower.

Reflection is an exploration and explanation of events. It is not reporting the events or the facts. Reflection includes your thoughts and feelings, which many students who are used to data-based studies will find strange. Reflection must be honest for it to work. It is very difficult, if not impossible, to work through an issue or problem when you have bad data. That is what it is like when you are reflecting without being honest with yourself. That lack of honestly gives you incomplete or incorrect data. Reflection should also be quite specific and not vague. You need to write about your exact situation at that time, not general accounts. For example, generally you would reflect on a specific problem, including the details of what happened on that occasion, rather than reflecting on having problems. Eventually, after a number of different reflections on different problems, you might put your findings together and draw some higher order issues. However, higher order understanding, or general understanding, comes from the individual reflections providing your insight first.

Employers Care About Reflection
The power of reflection as an important professional skill is widely recognised by employers. International consulting house PricewaterhouseCoopers list questions like "Do you learn from your experiences and take time to develop your personal approach to work?"[9] on their "How to land a job" page. Skillyouneed.com has an extensive page on reflection. For-profit career development companies advise that "reflection-in-action (at the time of the event) is a hallmark of an experienced professional".[10] And universities often feature information about what reflective skills are on their career pages. So let us learn about the nuts and bolts of reflection by looking at two formats for reflection.

Gibbs Reflective Cycle

Gibbs reflective cycle[11] breaks down thinking about an event into six portions: Description, Feelings, Evaluation, Analysis, Conclusion and Action Plan (Figure 2.8). Each of those portions has additional prompts (in dark blue) that can help you explore that portion. Once you get very good at reflection, you will find yourself running through these prompts either consciously or even without thinking about it. You will even find yourself reflecting while something is underway (reflection-in-action). For now, we will work on writing reflective pieces to help you develop that high level of reflective skill.

> *"Reflective practice is considered a key skill in many professions as it allows professionals to improve their practice by allowing them to become more self-aware, identify further educational needs, and monitor their own professional practice."[12]*

Common mistakes when writing any reflection (*i.e.*, what reflection is not):

- **Broad/vague statements** – reflection is most powerful when you look at specific examples. From a range of specific examples you might then draw generalities.
- **Too much description** – reflections need to be around 20% description but not much more. If you have much more than 20% description you won't have room for your reflection, and it will look like a report.
- **No learning or forward application** – make sure you include a plan or steps forward. This does not have to be a grand plan, just what is next to improve, why and how. What now?
- **Some external validation but no personal evaluation** – include what you think or feel and why, not what the teacher, professor or boss told you. Reflections are personal. Other people's impressions of you are often great feedback but reflections are a time to focus on your thoughts.
- **Low on honesty** – have you been honest? Is there anything you have not fully expressed because it is emotionally uncomfortable? Reflections do not need to make you look good. You might write down your most honest and sensitive reflections but keep them private until you are ready.

What, So What, and Now What (Boud)?

There is not a lot of difference between the two reflection schemes presented in this chapter. In the What, So What? scheme (Figure 2.9), Describe and Feelings are combined, as are Analyse and Evaluate. Some

Figure 2.8   The key elements of reflective writing: The Gibbs reflective cycle.[11]

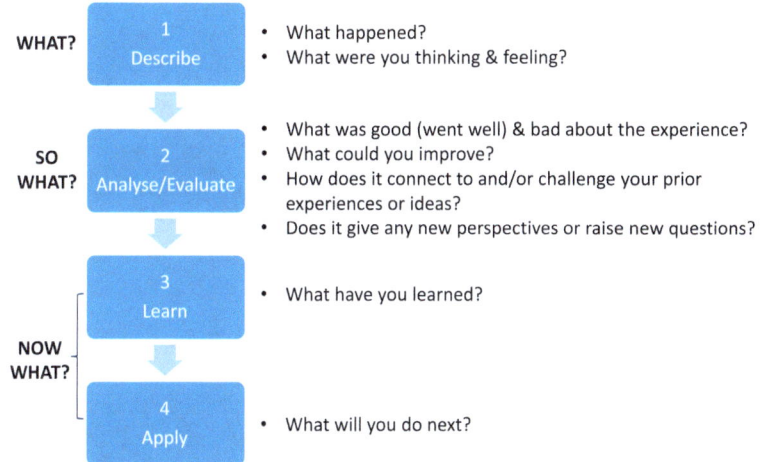

Figure 2.9   The What, So What? reflection scheme.[8]

people like this approach because they feel it is unnatural to talk about what happened separately to what they were thinking or feeling. Or they want to analyse and evaluate at the same time. The difference does not matter. It is just a personal preference. So, try both and see how you go.

Ways to Start Reflective Sentences

Most STEM students (secondary school and university) are taught to avoid the use of "I" when writing as scientists. The use of "I" is encouraged in reflective writing, and you will find it hard to write a good reflection (if not impossible) while avoiding the use of "I". If you are having trouble reconciling that fact, then think about reflections as being assignments or reports about yourself. You are the data/experiment. Your thoughts, feelings, and reactions are inputs, and your plans and further questions are outputs of your work. Because of this, if you do not fully include yourself it is really very hard to write a reflection. So, know that it is 100% OK to use "I" in a reflection and in fact we want you to.

Some examples of ways that you might start a reflective sentence with "I" are: I

Think	Remembered	Realised	Wanted
Wondered	Planned	Learnt	Concluded

If you get stuck starting a sentence while learning reflective writing, see if you can start the idea with one of the phrases in Figure 2.10 to kick-start your writing.

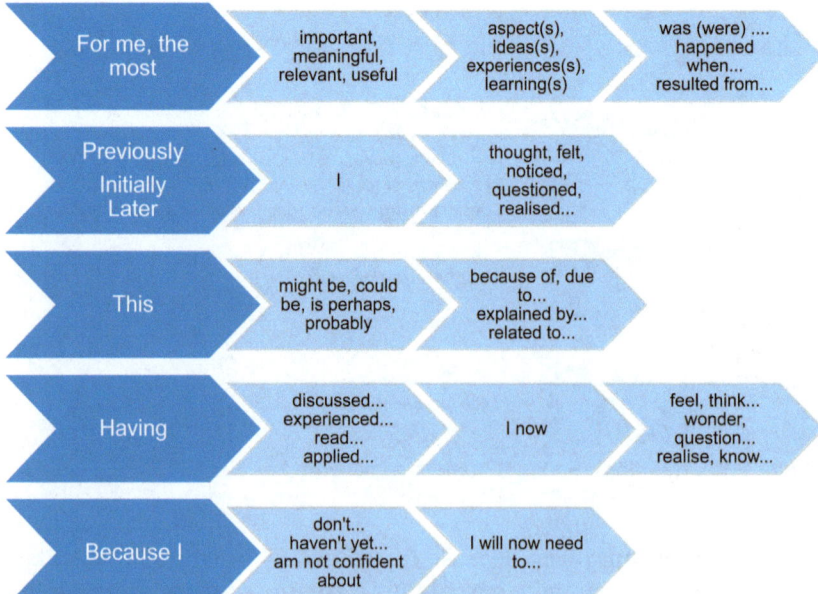

Figure 2.10   Some useful phrases to start reflective writing as suggested by the Australian Library and Information Association (ALIA).[13]

Reflection can help you identify what went well and what did not, what worked, and why. Or conversely, what did not work and why. By asking these what/how/why questions, you start to get an understanding of what to do next time to improve. Concluding what happened and why leads you to what to try next time. Continual reflection can lead to continual improvement in a system or situation until you reach a point you are happy with. Try Activity 2.7 to explore what a good or bad reflection looks like and why before you move on to Activity 2.8 and try your own first formal reflection. For both activities, it can be really helpful to talk to others about their thoughts. You might like to see if there are some others in your extended circle who want to learn more about reflection, and you can work together to learn faster and with greater clarity as you will gain different insights from each member of the group.

Activity 2.7: Analyse A Reflection for Its Strengths and Weaknesses

Activity goal: Improve your ability to understand what makes a good written reflection.

Purpose and benefit: This activity will help you get better at understanding how to put together a reflection, including things to avoid doing that take from your reflection.

Activity steps:

1. Following this activity, there are three reflections loosely based on student reflections in ref. 14 and 15. They each have their strengths and weaknesses. Pick one and comment on the strengths and weaknesses of the reflection in each of the six portions of Gibb's reflective cycle in Figure 2.8.

2. Look through the summary of common mistakes. Which of these mistakes (if any) are included? Detail the section it is included in and include mistakes several times if they happen in several places.

3. Check also for whether the right type of language is being used in the reflections? Do they put themselves at the centre of the writing and use "I" a lot? Is it personal or impersonal language?

4. Now that you have a list of errors, what would you recommend the writer do to improve the section? Write at least 40–50 words for each point to ensure you are fully detailing the mistake. Bullet points save time initially but often mean some thoughts and insights are lost. Given you might revisit this work overtime, it is particularly valuable to fully detail your thoughts.

Continued

Activity 2.7: Analyse A Reflection for Its Strengths and Weaknesses – (*Continued*)

5. Select a different reflection and work through the steps again, but this time use the second reflection scheme in Figure 2.9, the "What, So What?" scheme. This is a similar scheme, but you might prefer one over the other.

6. If you are working with someone else, now is a good time to share notes and talk about the example reflections to help understand reflections better. This includes learning from the insights of others as we all see and understand different points at different times when we learn.

Takeaway: Hopefully you now see how the "bad habits" of reflective writing lower the quality of the writing and can lead to less successful reflective writing. For example, not including a plan or not focusing on a particular issue but being very vague can lead to very unsatisfying reflections.

Reflection 1

Reflection on analysis skill: I have done well in analysis/problem solving. Reviewing my self-assessments, peer assessments, and feedback, I can see I've done well in this area. In classes I have received feedback on all my assessments that state that I carried out careful analysis of the topics and was able to articulate a valid and reasonable answer to the question.

My peers and instructors have mentioned that the responses are clear and logical, and patterns are well identified. It was also stated that I used referencing and the theory well.

I feel like I have the ability to analyse issues to solve problems well and that I use good practice in selecting, evaluating and then using resources. I am confident that I can come up with logical answers and solve problems in a logical manner.

Reflection 2

Reflection on field work: The field notes were written by hand on my note pad. They consisted of jotted notes and thoughts that I was having while I was doing the field work, including things I wasn't sure I needed to record but might be useful. I tried to make my observations relevant to the research questions.

I found the notetaking process itself helpful, and a bit calming. It ensured that I listened carefully and thought about what I was hearing. Some of the information I recorded wasn't useful but much of it was. I'm not sure I know how to do the note taking differently next time. I worry the notes were not a direct transcription of what the subjects said but consisted of interesting information that might not be the most valuable. I didn't have time to transcribe a direct quote so I just took notes, and I may have changed the meaning. Plus, when the speaker was talking about very technical issues, I couldn't always capture it correctly.

I will ask if I can bring a tape recorder next time. I know some people don't want to be recorded but at least if I had one, I could have used it.

Reflection 3

Reflections on starting an internship: As a biology student, I did not have many opportunities to be exposed to the corporate world. [...] the only jobs I thought about were things like a lab researcher. Maybe something in food safety. Anything to do with the business world felt quite unknown.

When the internship program started I was confused about how I might fit into this world. It seemed like most of the people signing up for the internship knew what to do and how to fit in. They came from disciplines that I associate with business, like finance and law. But then towards the end of the day I realised that maybe they weren't learning that much. I was learning from everything and I felt like some of them were not really taking advantage of the opportunity. I was galvanised to learn as much as I could.

After the internship (8 weeks), the idea of working in that sort of environment didn't feel uncomfortable and unknown. Now I knew what opportunities there might be in such a workplace. I could see myself working in such a workplace and how there might also be a career path. For example, I could work in a group with mixed specialists and it would be my role to bring in the biology knowledge to the project through research that I would understand and others might not be able to interpret. Or maybe I could work in environmental testing, some of those companies that carry out such testing are quite large and I didn't realise that. I'm very grateful that I had this chance to discover a type of workplace that I really didn't even know existed before.

Activity 2.8: Write Your First Structured Reflection – What Do You Want To Get Out of This Book?

Activity goal: Reflect on what you want to get out of this book.

Purpose and benefit: You will better understand what you hope to get out of this book, as well as your motivations behind that and how your motivations interact with other parts of your life. Some of these points might have been obvious to you before, but by reflecting more deeply you will have a better understanding of your motivation and be in a better position to make an action plan.

Activity steps:

1. Following either reflective scheme (Figure 2.8 or Figure 2.9), write 50–150 words on each of the portions of the reflective cycle. Double the amount you write for portions where scheme sections are combined in Figure 2.9. Remember to use the prompts in each figure (those in dark blue), and think around the prompts. Write in full sentences, avoiding bullet points other than when you are planning/drafting at the start of the activity. Try to make sure each thought gets finished. If you have a lot to write on one portion and go over 150 words, that is fine. This is just an average guide.

2. Check the common mistakes list to see which of these points you have accidently included. It is very common for most, if not all, of these "bad habits" to feature in first drafts of early reflective practice. But knowing these are the common errors makes it easier to check to see if you included them and work out strategies for avoiding them. Try some of the structures from Figure 2.10 to try to reword for a stronger reflection.

3. Work on your reflection for up to another 20 minutes or so, trying to minimise common mistakes and using Figure 2.10 for inspiration if you can't get started. Make sure you have a good plan with some approximate dates for those steps to be put into action. Put these dates in your calendar to remind yourself so they do not get forgotten when life gets busy.

4. If you have someone who can review this piece of writing, now would be a good time to show them and ask them if they thought you still included any of those bad habits of reflective practice.

5. Write out two key points that you learnt in this process and keep them in mind for when you are writing reflectively again.

Takeaway: You will have a plan of how to get what you want from reading this book and doing the activities it contains. For those of you who are towards the

**Activity 2.8: Write Your First Structured Reflection –
What Do You Want To Get Out of This Book?
– (Continued)**

start of your journey, this might be a career or skills exploration question or set of questions. For example, you might want to more fully understand your strengths and how best to find a career that takes advantage of them while working on a couple of weaknesses. For others, you will have a clear path forward but performing a structured reflection writing task will usually help tease out some details and help you formulate a plan or plans. If/when you revisit this section in a few months, or in a few years, the insights you gain and the plan(s) you make will be completely different, which is totally normal.

Time to Reflect Regularly to Build Reflection as a Transferable Skill

Now that you have a much more developed understanding of reflection, where might that take you? Experts (not me) recommend that people set aside regular time to reflect. At the start, this might be with a reflection scheme, but over many practices people find that they often do not need the structure or maybe they refer to the structure when they are a bit stuck and need some additional prompting. After a lot of practice, those that reflect a lot find they do it while the event is still happening around them. For example, a senior technician recognises that someone else on the team is giving unclear answers and last time this happened the instrument was repaired incorrectly. They take a moment to think and recall that the last time this happened, the same person had missed some of the information due to poor hearing but did not want to admit what they felt was a personal weakness. So, they decide to take ten minutes of their day to have a chat with their fellow technician about how the repair is going, ensuring that they talk without background noise so that there is no hearing issue this time. After talking about the repair in a friendly manner, the senior technician is reassured that the other teammate is now talking clearly about the repair and knows what to do. This is reflection in action and an excellent professionalism skill. Also, a great general life skill.

This simple example of reflection in action is also a great example of understanding your teammates, understanding we all have weaknesses, and that there are reasons people don't always talk about them. Plus, by understanding those we work with, we can find ways to politely ensure nothing goes wrong. Or at least minimise the risks.

In the future this person might even reflect further and plan to try to make meetings quieter or politely ask that particular technician to sit by the front to help.

Reflection Can Help Me With Everything?

Reflection probably cannot help you with all your career related questions and problems. However, it will come fairly close. If you are stuck working something out or making a decision, reflect on the big issues that are relevant. If you are struggling to look at the issue, take out the step-by-step instructions for reflection and just focus on one issue at a time, remembering the sentence starters. These reflections might just turn up more questions to ask, but that's OK. As STEM professionals, we often find questions behind questions. The further you progress working through the issue, the closer you get to knowing the full answer. In the meantime, you will get to a better understanding of the issue each time you work on it.

Remember that you are not an island and tough issues are best worked on with some outside input. That might be input directly on your reflective writing but will just as likely involve gathering more information and thinking about another's perspective. Or if you are lucky, you gather input from several people (either at once or over time) and you can start to put together a solid answer for your questions or problems.

For example, Chen has been trying to work out where they want to go with their career. They like engineering and health and business and coding. That means they are not like many of the other students they see in class. They have really broad interests. They were lucky enough to take an engineering degree where they could do all of those topics although less of some subjects and more of others. However, now they do not know where to go, and the types of jobs they hear students and career professionals talking about do not seem like things they want to do. They are not excellent at any one of these areas of interest but are very good at all of them. Chen feels like there is no career that uses all these interests and is thinking of giving up trying to combine them.

> **Hint:** There is always somewhere that you can use any combination of skills.

The first step is realising the issue and being honest about it. I've painted a clear picture of Chen's situation, but that clarity takes time to work out when it is you in a difficult position. Once you have a situation like

Chen's, it is outside information that will help. Thinking harder probably will not help. So be sure to balance reflective writing with a plan to learn more in general, but about your specific environment, too.

Is There a Bad Side of Reflection? – Analysis Paralysis

We all likely know someone who, when a problem happens, does not stop thinking about the things that went wrong, what they could have done differently and how this makes them look bad personally. They likely look at the situation over and over and almost start to obsess over the situation that occurred. This can lead to a type of analysis paralysis. The stress of the bad experience is stopping the person from moving on. The stress they are experiencing is real and it will stop them from trying new things.

While this might look like a negative side of reflection, it really is not reflection at all. Remember, reflection ends with a conclusion and a plan of how to improve, move on, or change for some positive outcome. Someone stuck on worrying about a problem can be helped by reflection. They are not over reflecting.

What Next?

Now that you have learnt about transferable skills and reflection, it is time for you to learn more about the workplace and the process of applying for jobs. As you work through these chapters, be sure to return to a focus on developing, understanding and articulating your transferable skills, and in using reflection to help you understand what you want and what motivates you. Both transferable skills and reflection weave in and out of all the following chapters. So set up your reflection practice schedule (Activity 2.9), print out your plan for improving a transferable skill (Activity 2.9) and enjoy using your newly acquired understanding of them to build your employability as you read about the different topics that help you prepare for your next job.

Activity 2.9: A Career Challenge That Really Worries You

Activity goal: Reflect on a transferable skill that worries you.

Purpose and benefit: To practice using your reflection on a more difficult topic in order to help you better understand that topic.

Continued

Activity 2.9: A Career Challenge That Really Worries You – (*Continued*)

Activity steps:

1. Pick a transferable skill you are worried about trying to improve or do not think you will be able to improve.

2. Select your reflective cycle (Figure 2.8 or Figure 2.9) and write a reflection by first addressing the prompts for each portion of the reflective cycle.

3. What did you find more or less difficult this time? Why do you think that was? Do you think more reflective writing will help?

4. Build a plan to write reflectively in a semi-regular manner. This might mean an hour each Sunday morning or 30 minutes on public transport 2–3 times a week. Checking in at the end of the semester/trimester and in mid-term breaks is a good idea too. However, you might want to build your reflection skills to start with by spending more than an hour or two every couple of months. Write a plan that might work for you and put it in your calendar to remind you if you forget.

5. Put another reminder in for 3 months' time to check on how you have been going with your plan. Is it working? Why? Why not? Hang on, you are reflecting again!

Take away: Some things are harder to reflect on. There is learning research that says that the best learning comes from challenging premises you have about yourself.[16] This can be hard work and that is OK.

References

1. Y. S. Banu and R. K. Angamuthu, *Shanlax Int. J. Manage.*, 2022, **10**(1), 9–12.
2. M. A. Hill, T. L. Overton, C. D. Thompson, R. R. Kitson and P. Coppo, *Chem. Educ. Res. Pract.*, 2019, **20**, 68–84.
3. J. M. Kiweu, D. Mulwa, J. Kinyili, P. Muriungi, R. Kimiti and J. Muola, *Eur. J. Educ. Stud.*, 2021, **8**(2), DOI: 10.46827/ejes.v8i2.3621
4. D. McGunagle and L. Zizka, *J. Aviat./Aerosp. Educ. Res.*, 2018, **27**, 59–76.
5. G. M. Rayner and T. Papakonstantinou, *J. Teach. Learn. Grad. Employability*, 2015, **6**, 110–125.
6. M. Sarkar, T. L. Overton, C. D. Thompson and G. Rayner, *Int. J. Innov. Sci. Math. Educ.*, 2016, **24**(3), 31–48.
7. M. Sarkar, T. Overton, C. Thompson and G. Rayner, *Int. J. Innov. Sci. Math. Educ.*, 2017, **25**(5), 21–27.
8. D. Boud, R. Keogh and D. Walker, *Reflection: Turning Experience into Learning*, RoutledgeFalmer Taylor & Francis Group, London and New York, 2013.

9. cnbc, https://www.cnbc.com/2018/06/13/how-to-land-a-job-at-pwc.html.
10. Bradley Shearer Our Stories, updated November 29, 2020, https://www.prota-gion.com/stories/professional-reflection, accessed on October 2, 2022.
11. G. Gibbs, *Learning by Doing: A Guide to Teaching and Learning Methods*, Oxford Polytechnic, 1988.
12. L. Pretorius and A. Ford, Reflection for Learning: Teaching Reflective Practice at the Beginning of University Study, *Int. J. Teach. Learn. High. Educ.*, 2016, **28**(2), 241–253.
13. ALIA Reflective practice writing guide 2024, https://www.alia.org.au/com-mon/Uploaded%20files/ALIA-Docs/Reflective-Practice-Vocabulary.pdf,accessed on October 16, 2023.
14. W. H. Rickards and M. E. Diez, *et al.*, Learning, reflection, and electronic port-folios: Stepping toward an assessment practice, *J. Gen. Educ.*, 2008, **57**(1), 31–50.
15. University of New South Wales, www.student.UNSW.edu.au/examples-reflec-tive-writing, accessed on August 13, 2023, page last updated 21st June, 2023.
16. J. Mezirow, Understanding transformation theory, *Adult Educ. Q.*, 1994, **44**(4), 222–233.

Commercial Awareness

REBECCA YEE

GPA Engineering, Australia

Introduction to Commercial Awareness

If you were born 1000 years ago, you likely lived off the land – gathering fruit, catching fish, setting traps – everyone chipping in and doing a bit of everything. It was only as our social groups grew in number that individuals began to specialise so much. Skills and labour could then be traded for things that you couldn't make or get yourself. Today, our modern lifestyle and conveniences are the result of many, many specialised individuals working in different areas of business or production.

Commercial awareness is understanding your employer's business, products, or services, and how they fit in today's markets. This can help you frame how your job and role work within the organisation.

Markets have long served as hubs for commercial activity, dating back to ancient civilisations. If you wanted something that you couldn't grow or make yourself, you needed to trade. In smaller villages, direct bartering (I'll give you this goat in exchange for 3 of your chickens) worked well enough. As communities grew, trading hubs became known as markets. A market can be a physical or virtual place where buyers and sellers come together to exchange goods, services, and information. Markets facilitate the efficient allocation of resources by allowing buyers and sellers to find each other and negotiate prices.

In a market, buyers have the opportunity to compare prices and quality across different sellers, while sellers have the opportunity to reach a larger pool of potential customers. This competition among buyers and sellers aims to price goods and services based on supply and demand.

Markets (Figure 3.1) can take many different forms, from traditional physical marketplaces like bazaars or farmers' markets, to online marketplaces. Markets play an important role in economic activity by facilitating trade and enabling specialisation of labour and skills.

Figure 3.1 Markets can vary a lot in shape and size, but these are essentially places (physical or virtual) where people meet to trade; photo by Roberto Carlos Román Don on Unsplash.

Markets have various names in different cultures and regions (Agora, Forum, Bazaar, Souk, Hui, Marts, Plaza, Tianguis, Fairs, Trading Posts). Some would last for days, weeks or only be held occasionally during the year. These places are not only for commercial activity and trading but are essential aspects of society.

Why Is It Important to Be Commercially Aware?

We all need to eat. Shelter and internet are nice too. Most of us need a financial income. Today, a stable, potentially ever-increasing source of money is tied to a job or series of jobs – also known as a career. But many new graduates will leap into the job market and apply for anything and everything they think may be relevant. Being commercially aware will help you to better focus your efforts. This chapter will give you the tools to recognise key commercial concepts and structures. This will guide you in understanding where your skills and experience may best fit.

In the modern working environment, employers often state that they look for the following skills (see Chapter 2 for more on transferrable skills):

- critical thinking
- emotional intelligence in the way you communicate, work with, and treat other people
- complex problem solving
- creativity
- time management

- adaptability – technology and cultures are changing around the world. As COVID-19 and other pandemics have taught us, being able to adjust around unexpected changes is a very useful skill
- reliability.

While you may recognise these qualities in yourself, the hardest part for many graduates is how to 'demonstrate' these skills to employers. It is one thing to **claim** that you are a critical thinker, quite another challenge to **explain** this in an interview. The best way to paint a picture of your abilities is to give specific examples of **how** you have used such skills. Being commercially aware will help you to **recognise** where you have used certain skills in your everyday life, often in things you already do. For example, perhaps you volunteer every week at a student drop-in tutoring class. This could demonstrate your time management skills, reliability, as well as adaptability if you needed to work around certain student issues.

Being commercially aware is not about knowing the details of the global economy (impossible really), but rather just having a broader appreciation of how jobs and organisations interact to make products and offer services. This will help you consider the context of your work and the types of jobs that could use your qualifications.

This chapter will help you start the conversation. **Ask your friends about their part-time jobs.** Learn how different types of employers have different perks. Start thinking about what type of employer would suit you. What type of company or benefits are important to you? Do you care about free parking if you don't have a car? Is it really just about the money or would you prefer a job that allowed you flexible hours or a mission that you were passionate about?

You are more likely to find a job that you enjoy if you broaden your awareness of what is out there. **Start subscribing to industry newsletters or networking events.** Learn about industries, groups, or organisations that could use your talents. The more you learn, the more you will be able to see the bigger picture.

> *"It is hard to see the forest for all the trees."*
>
> – John Heyward, *ca.* 1546.

This proverb essentially means it is challenging to see the bigger picture when you are in amongst it – a part of the system itself. Even more so when this system is constantly changing. New job titles are created each year with the growth of new industries and markets. Even market systems continually fluctuate as politics and society shifts.

As you try to find your way through the forest, it would be almost impossible to proceed without a map or some greater understanding of the wider area. Charting your career may sometimes feel like a trek through the jungle without a map. Without a bigger picture view, it's hard to know which direction will suit you best, what career options or pathways there may be, and where to find help or guidance. Let's hear from Jim (Scenario 3.1) and Jing (Scenario 3.2) and learn how people's commercial awareness journeys differ.

Many students, like Jim, only become aware of opportunities when they speak with someone about it. Others, like Jing (Scenario 3.2), had the daily experience of planning for their future. Hopefully this chapter makes you aware of how our economy and commercial systems work, how you can navigate the employment structures that have developed over 200 years, and how to start the conversation to learn where you fit.

Being commercially aware will make the transition to working life much smoother. It helps if you can see how your role fits into an organisation. How does your job contribute to the companies' operations? Commercial awareness will help you recognise more about how businesses operate. More importantly, commercial awareness will help you see where you can play a part.

Scenario 3.1   Jim's Journey to be a Fish Biologist

Jim was born in Port Lincoln, South Australia. His family have traditionally been fishers, his father owned and operated his own boat as part of a contract for a local fish producer. But Jim doesn't want to be a fisherman. When he was a child, he always wondered where the salmon went during the winter. As he grew older, he learnt about migration, which made him curious about animal behaviour, which then led him to study flocking behaviour. Jim became an expert in the computer models and software used in these studies and was then hired by an architecture firm to use the same principle in designing urban centres around the way people move.

Jim's parents were happy that Jim was pursuing his own interests and wanted to support him, but they had left school early themselves to work on the family boat or help with the house. They had no idea about landscape design or computer modelling or even how to get a higher education. But in Grade 3, a teacher noticed his frequent questions about where the fish went and showed him how to use the school library computer to look up the routes of local species. In high school, while having dinner at a friend's house, their parent, who worked at the local power plant, told Jim about how he studied science at university to become a mechanical engineer. This conversation was the first time Jim became aware of scientific careers.

Scenario 3.2   Using an IT Degree to Help the Family Business

Jing was born in Fuzhou, China. His father had built a successful stationery and printing business and, ever since Jing was born, it was his passion to pass on this business to his son. As Jing started to learn basic maths, his father would give him examples related to the business, such as how many pens you needed to sell to make a profit. Jing grew up with a direct awareness of commerce and economics but, perhaps because of his father's financial success, he was always more interested in the newest technology rather than making more sales.

Jing excelled at his high school accounting classes. He knew he would always take over the family business but wanted to try something different for university. He decided to study computer science, where he learnt about robotics, automation, and new ways to analyse data. When he graduated and took over the stationery business, Jing realised he could apply his love of technology to also improve productivity and help such a traditional paper business grow with the digital generation.

"The whole is greater than the sum of its parts."

Another important aspect of commercial awareness is recognising that the system is not intended to be one-way. Just like a forest benefits all organisms by distributing nutrients and providing buffers for hard times like droughts, organisations can also provide benefits to employees, not just through financial income but career-wise. Employees often gain opportunities to impact the world in ways they could never do by themselves. For example, working for a company with offices around the world would make it much easier to travel and live internationally (if that is your dream). Some companies support their staff during career breaks to raise children and return to the same type of role. Other organisations may pay for membership to industry groups or networking events. Many opportunities to learn and grow professionally will come directly through your employer, which may be of particular importance if you want your career to align with a purpose. After all, we can do more as a group than any individual.

What Is Commerce?

Think about how goods are exchanged, the way we pay for services by time or by task. Commerce includes the process of getting goods from a manufacturer to the customer and everything in between, also known as a supply chain. This can also include the research time and equipment needed to come up with an invention or product.

Modern commerce was initiated by the Industrial Revolution. Previously, goods were produced by hand in small quantities, and trade was primarily conducted locally. In the 1760s, the Industrial Revolution used centralised sources of dense energy, like crude oil and coal. These powered machines like the steam engine. New technologies and manufacturing processes were developed, which allowed goods to be produced more efficiently and in greater quantities, leading to the rise of large-scale industrial production and mass manufacturing. Processes became automated. Products were manufactured that were made by teams of people and sold by companies.

This increase in production, combined with improvements in transportation and communication, enabled goods to be transported more quickly and cheaply over longer distances, allowing for greater trade and commercial activity.

The industrial revolution also had a significant impact on the development of modern business practices. It led to the rise of large corporations and the development of more formalised business structures and processes, such as profit-driven supply chains, project management to improve efficiencies, marketing, finance, and intellectual property to protect valuable assets. In this chapter, we explore these topics to give you a wider perspective of how your skills and career can fit into the bigger picture. The aim is to give you more insight into the opportunities that exist around you.

Over the last 200 years, technological inventions and several industrial revolutions have shaped our world. It wasn't just physical goods that moved along trade routes. Exposed to new cultures, ideas, and behaviours, Eurocentric science and philosophy also spread. Concepts of our modern civilisation – running water, shared infrastructure, systems of money, debt records, accounting, and even the concept of taxes or social justice were developed. STEM jobs play a key role in nearly every commercial and technological development.

Forget About Your Discipline

The idea of compulsory schooling is a relatively novel phenomenon. Historically learning was done in communities as part of everyday life. There were systems of protecting and preserving knowledge, traditions as to who learnt what and why. For example, needing to learn X and Y before Z. There was no need to read and write for the vast majority of people. Post the industrial revolution, workers started to need more skills like reading and writing so structured schooling was a must. Science, maths, history, geography. Taken further, an interest in science could then lead to further advanced study of biology, chemistry, physics, anatomy, or machine learning (summary of learning fields in Table 3.1).

Table 3.1 One version of the main academic disciplines and some of their branches.

Discipline	Branch examples
Business	Accounting, economics, finance, management, marketing
Humanities	Art, history, languages, literature, music, philosophy, religion, theatre
Natural and Applied Sciences	Biology, chemistry, computer science, engineering, geology, mathematics, physics, medicine
Social Sciences	Anthropology, education, geography, law, political science, psychology, sociology

This is the easiest way to break down our educational process. Start with a broad, simple topic, and then add more detail to that topic. This is the most systematic way to disperse information (aka 'teaching'). But have you ever thought how such compartmentalisation of information translates to a day in a job? You don't really get paid just for what you know. Even winners of game shows have to compete and perform for an audience.

Many people who focus on becoming experts in their discipline and lack commercial awareness are called 'book smart'. They may have a solid grasp on a particular field or topic of education but are not aware of how it applies or is used. They may not even recognise jobs that use such information. Is Chemistry only used by Chemists? Who even works with machine learning or drones? How can you even find jobs that involve data analytics?

For years you have been trained with the mindset that information is learned discipline-by-discipline. As you transition from this student-learning perspective into a period of life where you are expected to contribute your skills to society, it can be frustrating and confusing without knowing how you fit in. Commercial awareness can help you see how disciplines can be applied to the working world. A good student doesn't necessarily mean a successful career. A student with great memorisation and cramming abilities may not succeed in getting a good job at all. Different skills are needed to compete in the job market. For example, in-depth knowledge is rarely useful but being a quick learner is often highly valued in many industries.

So, forget about your discipline. Think industrially. Start thinking commercially about how industries use different disciplines. How do these disciplines contribute together? What industries use your particular choice of study? What is it like working in those industries? Who else could you talk to about careers in these fields?

Activity 3.1: Thinking Beyond STEM Disciplines

Activity goal: Identify multi-disciplinary products and services.

Purpose and benefit: Expand your commercial awareness of STEM-related products and services.

Activity steps:

1. List 3 products or services that you can see around you that could be impacted by a STEM discipline. *E.g.* your pen, the building next door, the café you are currently sitting in, *etc.* But pick three new ones, not these!

2. For each item, consider what STEM topics are involved in that product or service. For example, a café that sells food would likely need someone with health and safety knowledge (and practice).

3. Consider what specific STEM disciplines would be necessary to complete that product or service and get it to you. *E.g.* A global supply chain sits around every cup of coffee. In addition to agriculture and processing the raw bean, there are also food handling practices at many stages using chemistry and physics.

4. Write down a specific company that sells the products or services you have identified.

5. Think about the difference between specific STEM-related knowledge and what companies trade in those products or services.

6. Think about what jobs and roles make up that company.

7. Repeat this activity every few months as you become more aware of products that use STEM skills.

Takeaway: Even when it isn't obvious, STEM training and careers are behind most everyday products. By being more conscious about this, you will start to better appreciate where you can and might want to use your skills.

The Global Economy

Global Opportunities

People have often moved for better opportunities. We followed migrating herds, travelled for seasonal harvests, changed communities for adventure, love, or new jobs. But this was all done at the pace of walking or, if you were lucky, how fast and far your horse (or camel or llama or local beast of burden) could roam. Towns established during this era were located within a day's walk of each other.

The Industrial Revolution, the use of machines, especially the invention of cars and planes, drastically changed the scale of human civilisation (think about the geographic layout of pre-industrial and post-industrial cities). We can now travel across the world with a few mouse clicks and a carry-on bag.

To gain some perspective on this concept, have a look at a world map. You can **technically** walk from Cape Town in South Africa to Beijing, China. This 21 970 km stroll would take you 4450 hours of continuous walking. Assuming you only walked during daylight hours (say an average of 8 hours a day), this would take you 556 days or about 1 and a half years (assuming you don't get lost either). Or you could take a 20 hour flight watching a few movies with meal service.

Globalisation refers to the increasing integration of economies around the world. And such advancements in fast and easy travel have led to increased globalisation (as well as many developments around regulations and tariffs). When my grandparents migrated to Australia by boat, the journey took several months, and the trip was expected to be one-way only. Now, it is fairly standard that higher education requires travelling away from home for most students. Especially for STEM graduates, there is an expectation that you travel – for more specific education or jobs, to field trips across the country, or even international conferences.

Science is widely recognised as an international system (think about the International Standard of Units). It is important to recognise that STEM jobs are global and that you may want to move to pursue your most desired career path.

Activity 3.2: What Global Opportunities Are There With STEM?

Activity goal: Identify multi-disciplinary products and services.

Purpose and benefit: Expand your commercial awareness of STEM-related products and services.

Activity 3.2: What Global Opportunities Are There With STEM? – (*Continued*)

Activity steps:

1. Pick any country.

2. Search any online search engine for a job in this country. For this exercise, this doesn't have to be realistic or even tied to your degree, just pick something that is interesting to you.

3. Have a look at a few job descriptions to see what type of requirements or qualifications are needed. For example, is it required to be a citizen or permanent resident?

4. What surprises were there? Try to list two positives and two negatives. What have you learnt about working in different places or in different environments? Can you identify explanations for any surprises you had? For example, wages varying a lot due to cost-of-living variations from region to region, high taxes eating into what seem like high wages, or how much benefits vary.

5. Repeat steps 1–4, two more times. Try different countries if that is relevant to you or you can try different industries or positions. If you aren't sure where you will work but know the country you want to work in, try three different major cities to start to understand the differences.

6. Put time in your diary to repeat this activity a few more times to slowly grow your knowledge of the environment for different regions, roles and industries.

Takeaway: Jobs and commercial awareness interact on many levels. Getting a base line understanding of where roles that might be of interest to you are and how they vary is a good start in understanding this interaction.

Think Global, Act Local

More than just an environmental slogan, it is important to remember that while the modern communication and globalisation makes the world seem smaller, there are distinct and unique aspects of your local economy. Your local businesses, supply chains, how your food gets to you, how your electricity is generated – even in this global era, these commercial activities are unique to your location. Even at a small scale, think about what shops you can reach within a 10 minute walk or drive. Can you access public transport easily? What kind of takeaway food options do you have? Is your electricity powered locally or generated by imported fuels?

I am writing this book on a computer powered by electricity (thank goodness for Ctrl + Z). I am drinking a cup of tea made from water heated by electricity. Even that water could only reach me using huge pumps powered by electricity. Life would be very different without electric power. We will use electricity usage and cost in this next activity as an example of how something as central to our lives as electricity has a commercial element. The cost of your electricity is based on your local electricity grid and the type of power sources that supply it. The method of generating electricity, the cost of the fuel, equipment, amount of manpower to operate (labour costs), even the varying cost of transmission lines (how far are you from the power station), is what affects your electricity costs. We will also consider the power source.

Activity 3.3: Review Your Electricity Bill

Activity goal: Identify how every day costs relate to STEM units.

Purpose and benefit: Expand your awareness of commercial systems that use STEM.

Activity steps:

1. Find a previous electricity invoice for your home – ask your parents, landlord or search your email inbox or bank debits. Are you responsible for paying your electric bill or is this the first time you have really looked at such a universal invoice? This time, have a closer look.

2. Answer the following:
   - How many days does the invoice charge cover? You may have to do some maths from the date range.
   - What is the total cost ($) on the invoice?
   - How much does your electricity cost per day?
   - Does the invoice state how much total energy you used (kW h)?
   - Does the invoice state how much is the cost per kW h?
   - Are there any other charges, such as connection fees?

3. Some of this information may only be displayed on invoices in certain countries. If you cannot find such data, try to find the website of your electricity provider to get this information.

4. What powers your local electricity grid (e.g., solar, coal, natural gas)? This may be stated on your electricity bill, on your electricity provider's webpage, or through a general search of your town's name.

Activity 3.3: Review Your Electricity Bill – (*Continued*)

5. What have you learnt about energy and pricing doing this? Is energy costing you more or less than you thought? Do you think the daily provider fee is reasonable (if charged)?

Takeaway: This relatively simple commercial awareness activity has given you an example of the types of financial issues that STEM professionals might need to take into consideration. The price of energy factors into so many things: transport, which appliances we buy/use, when we upgrade systems, whether businesses make climate-wise decisions about energy sources, whether our product is cheaper or more expensive.

As a STEM graduate, it is useful to consider where you live and your local employment opportunities. For example, if you live in a city with lots of restaurants, it is probably going to be easy to get a job as a chef or waiter. Start thinking about where the jobs you want may be located. Where in the world is that particular industry thriving? Try to recall what prompted you to study STEM in the first place. Are there any issues or situations in your local community that could be improved using STEM? Are there any organisations that work in this area?

Activity 3.4: What Companies and Jobs Exist Locally?

Activity goal: Identify local commercial activities related to STEM.

Purpose and benefit: Expand your awareness of employment opportunities (or lack of) nearby.

Activity steps:

1. Make a list of the types of shops or businesses you see when you walk around your town or city. If you can't remember, have a walk around the local shopping centre, or business/innovation park, and note the types of companies located there.

Continued

Activity 3.4: What Companies and Jobs Exist Locally? – (*Continued*)

2. Tick off the main industries these businesses fall under:
- agriculture and food production
- construction
- entertainment
- finance and professional services (accounting, legal, consulting, insurance, *etc.*)
- health
- manufacturing
- mining
- tourism
- transport
- retail.

3. Visit an online job search engine and select the geographical filter to see what types of jobs are local to you. Search very broadly, jobs often don't fall under your degree or major title. Select three that might suit you and investigate by looking at the qualifications needed.

4. If there aren't any or aren't many roles that you might be interested, head to Activity 3.5.

Takeaway: A lot of STEM jobs are not that obvious. Doing research to find out what is in your local area is very important in helping you get to know the job market in your area. Following it over time will help you learn more and mean that you are able to spot trends that might help you.

Activity 3.5: How Global Is the STEM Job Market?

Activity goal: Identify job requirements for international STEM roles.

Purpose and benefit: Expand your awareness of standard employment expectations and consider international opportunities.

Activity steps:

1. Search online for any job that takes your interest that is NOT in your geographic location. It does not have to be realistic or relevant to your field of study. This is just an exercise to consider your options.

Activity 3.5: How Global Is the STEM Job Market?
– (Continued)

2. Review the job and note the following:
- Does the job advertisement or company website state their policy on working-from-home? Is it discouraged or allowed?
- Does the company have offices internationally? This may mean they are already set up to accommodate international staff.
- Can you be paid from one country and live in another? What does this mean for your income tax obligations? Who do you pay tax to?
- Are your qualifications relevant for your intended country of work?
- Do you require a visa or residency permit? Will your company support and help you in this process?
- Is the pay scale suitable for the cost of living in your intended location?

3. Select other job descriptions to review and gauge what the 'standard' expectations are for your field or geographic area of interest. For this, pick 5 variables that you will compare. Common ones to start with are things like wages and benefits, but there are no wrong or right variables to select.

4. The activity is just for collecting information, no need to make a plan. What three notable points have you discovered in your investigations?

Takeaway: It is often just as useful to see what else is out there, whether you are thinking of moving or not. By knowing what other areas are like we can evaluate the benefits of staying put and make an informed decision, whether that is to stay, to move, or to just do more research so you know more. This is an excellent activity to come back to as your needs change or your career thinking develops.

Global Challenges

Graduating in this post-pandemic era is another unknown. While I make claims to advise you of your career and economic opportunities, the global job market has undoubtedly changed in a few short years. International travel is no longer as convenient, although this could evolve in the future as well. Many live events and networking opportunities have even been permanently replaced by online sessions. Working-from-home has become a common phrase. Job advertisements now make note if the role is hybrid or flexible.

Work-life balance has become another hot topic. Many younger people recognise that financial wealth cannot buy a happy life. While some companies are encouraging their employees to return to the office, others may allow more flexibility to attract highly skilled workers. Some organisations don't even care where you are located at all. Does this mean you can work anywhere in the world?

On top of international and legal requirements, there are usually numerous cultural aspects to consider if you want to work in another country. These social considerations are a critical aspect of being culturally aware. Does your workplace have a strict hierarchy or are decisions made by consensus? Are you expected to call people by their surname or are nicknames OK? A broader perspective will help you to communicate better in new social circles. For example, in some countries it is considered rude to share how much money you make. While in others, it can be encouraged, and salaries may even be published online.

Sectors, Industries, Economies and Government Structures

Educational systems have shaped our thinking around fields of discipline. Art, science, history, physics, music, geography. These make topics easier to break down and learn. However, this mindset is where many graduates get stuck when looking for employment.

The commercial economy is not structured around topics of study. A new graduate who majored in chemistry may say they want a 'job in chemistry' but this is just not a thing. Workplaces are like a jig-saw puzzle (Figure 3.2), many pieces are needed, not just one discipline. A graduate who is more commercially aware would recognise that they should

Figure 3.2   It takes many intersecting disciplines to build our civilisation.

look for industries that *use* chemistry. This could include anything from car manufacturing (plenty of polymers, paints, sealants, solvents, and special materials are needed) to water utilities (important chemistry is needed to treat wastewater and make sure drinking water is safe for the public) to ice cream production (very important chemistry is needed to make good ice cream).

Most industries and companies want employees from a range of disciplines. Being aware of how these sectors are shaped will help you to see the broader opportunities where your discipline knowledge can be applied. You will also learn how your role interacts and works with others.

As a fresh graduate, there can be an uncanny urge to just get a job – any job. This seems the most obvious step to get your foot in the door, but many young professionals jump into a role without thinking about how the industry works or what career paths may be available within that industry or company. These graduates often become unsatisfied when they realise the lifestyle or company culture does not align with their values.

For example, a mechatronics engineer may be drawn to the defence industry just because it is the most obvious connection to working with robots and rockets. However, stressful deadlines, bureaucratic culture and frequent travel may not suit how they want to live. They may have been better off working for the local hospital that develops robotic prosthetics for amputees, where they can live near their family and feel more job satisfaction because they are directly working with the people they are helping. Commercial awareness helps students recognise opportunities that align with their career goals, even if they may be through 'non-traditional' employers. Another example is in the mining sector, where many environmental engineers are being snapped up so that companies can meet their license requirements. You can think about the sectors and how they interact to help you when looking for employment (Figure 3.3).

It is even more common that many students think about life-after-school as a choice between 'getting a job in industry' or 'staying in academia'. In this chapter, I hope to steer you away from this mentality. Instead, it can be more helpful to think of sectors around public (government) or private organisations, and to consider the services or products that they provide (Figure 3.4). There are also other types of organisations such as not-for-profit charities or research programs that may hire academics but are led and funded by private companies, or even co-funded by

Figure 3.3   Developing a career around an economic sector or industry can expand your awareness of suitable jobs and fields.

Figure 3.4   There are many different employers out there, they are grouped into six categories.

private companies and government groups. Many industries, especially in STEM fields, are closely integrated with academic institutions to innovate and solve technical challenges.

In this chapter, several categories of employers are outlined. These are not hard or even clear groupings and may not be complete. However, this will hopefully provide an introduction to organisations that may be commercially focused, or associations that have alternative missions or support roles. This will help you recognise opportunities for employment across a range of institutions.

Activity 3.6: Trace the Supply Chain of a Product

Activity goal: Identify the relevance of STEM topics in everyday products.

Purpose and benefit: Expand your awareness of supply chains and industrial sectors.

Activity steps:

1. Pick one of your favourite products. It can be a food, item of clothing, or something that entertains you.

2. Search online to find out:
   - Where is that product made? Are all the ingredients or parts sourced from the same place?
   - Does the company that sells the product also make it?
   - Lists the steps required to make and sell the product. Can you list which industry each step belongs to?

3. Select a second product in a completely different range of products and repeat 1–2.

4. What did you observe? Are the lengths and complexities of the systems surprising or have you learnt about supply chains elsewhere?

5. Are your examples more or less complicated than the Nutella® example discussed in the next section and summarised in Figure 3.5?

Takeaway: Products are much more complicated than we tend to assume. Thinking about the supply chains can help us better understand the STEM that is required, but also all the other elements that are required in their production. This can also help us appreciate how many STEM professionals work closely with other professionals and where that occurs, both geographically and in terms of the portion of the supply chain.

A popular product in some countries is Nutella®. Nutella® is made from sugar, vegetable oil, hazelnuts, milk powder, cocoa powder, emulsifiers and vanilla. These ingredients are sourced from many countries (see Figure 3.5) and touch on industries from agriculture, food processing, manufacturing, food safety, packaging, cold storage shipping, logistics, and the always-important marketing. It is owned by Ferrara, a private Italian company that is also the second largest chocolate producer in the world, with 18 factories globally. There are many STEM-related jobs that would work for this company, or

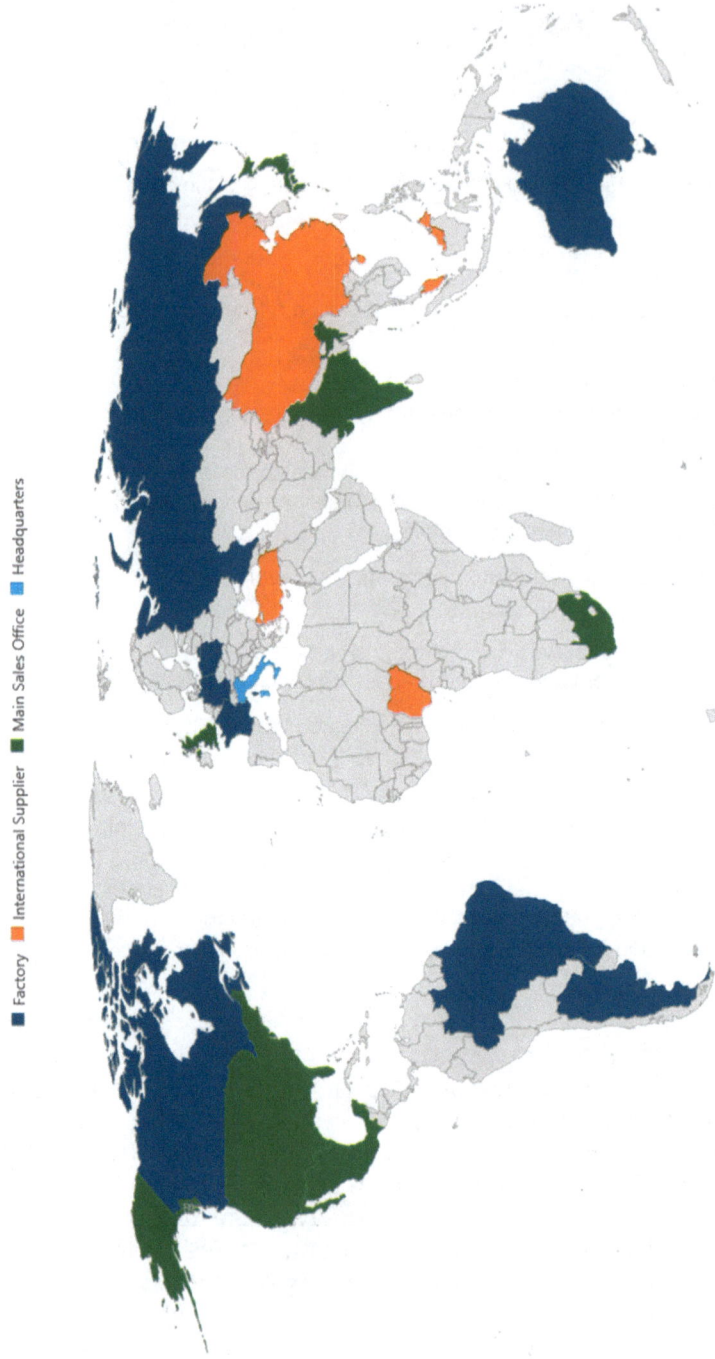

Figure 3.5 Geographic locations that contribute to Nutella®. Data from Australian Bureau of Statistics.

somewhere along the supply chain, to get a jar of Nutella® into your cupboard. You can see the important regions for the production of Nutella® in Figure 3.5.

The Rise of Private Business

Private companies are the first place most new job seekers look. These could be small businesses with one owner, such as a local construction company, or could be an international group owned by a family or board members.

Private businesses often have interesting histories and are worth exploring if you have identified companies that make or do something interesting to you. It is always worthwhile to research a company's culture to ensure it aligns with your own values.

Hiring processes can be very different for private organisations, for a large international corporation, or a small start-up company. Private company jobs, especially for smaller groups, can be shared just by word of mouth. There is usually no law that a job vacancy must be publicly advertised. It really is true that it is not **what** you know but **who** you know. Many people say this with a fatalistic sigh of exasperation – as if this is an unfair aspect of life – but it could be seen as an opportunity. I encourage you to speak to people about your career interests. You never know if your aunt's neighbour's grandson's girlfriend owns a company in your field!

You may know some privately owned companies such as IKEA, Nestle, and Aldi. Some of them have been around for nearly a century. You may not know exactly what they do but all of these private companies hire STEM graduates.

Publicly Listed Companies

Shares or stocks are used to describe the units of ownership of a company. A private partnership may see one owner having 80 shares while the other has 20. This typically means that profits (and losses) are divided on this basis. However, for public companies, these shares are traded on a stock exchange that anyone can access and buy into. This means that anyone in the world can own a part of that company. These owners are known as shareholders. This allows companies to raise more capital by selling shares to numerous investors. The companies invest those funds back into their business, and investors, ideally, earn a profit from their investment in those companies.

Activity 3.7: Identify Publicly Listed Companies

Activity goal: Find your local Stock Exchange.

Purpose and benefit: Develop awareness of industrial sectors and large corporations.

Activity steps:

1. Look up your 'Country Name + Stock Exchange'.

2. Find the list of companies. Do you recognise any brand names? Which ones?

3. Identify five companies listed. For each company, find the company's website and go to the careers page. This might take a few goes but most will have jobs/careers pages.

4. List any STEM jobs that you might qualify for. Be very expansive and read the descriptions if there are some you are unfamiliar with. Are there any STEM-related jobs advertised that you are interested in?

5. Explore the Stock Exchange website. Many have information to educate you further on how the market works.

Takeaways: The stock exchange will represent some of the biggest employers in your area so it can be useful to be able to identify large employers and determine what sort of graduates, and potentially students, that they employ.

Government Departments

Every government in the world employs STEM qualified employees. While many of these roles are not actively engaged in scientific work or research, the skills gained from such qualifications are highly sought after for government jobs. Departments and authorities, such as ministers or agency leads, are responsible for shared social issues such as health, education, and justice systems; infrastructure such as roads, railways, water pipelines and electricity grids; defence and security; and, of course, tax agencies. Even in non-scientific roles, critical thinking and data analysis are essential to support all of these important aspects of our civilisation.

For example, in Australia, over 8% of the population are employed by government departments.[1] That's one out of every 12 people.

"Being in government is very different to being in the private sector. Red tape, I guess, you've got a lot of different hurdles to what you do. In my other jobs, there are certain policies that you have to follow, procedures, and just the way they work, is a little bit different. There's lots of reviews and lots of clearance, that you need to get for work that you do, which I hadn't really experienced before. I was just like, oh, you go to your manager. If your manager is happy, then it goes on. But yeah, I suppose there are more considerations in a government job than I was expecting."

– Megan, science graduate with honours in science education. Started in a consulting graduate program; moved to policy in government.

And All the Others

When thinking of career options, there are several other organisational structures that employ STEM graduates that don't fit into these groups. This can include international government organisations, like the United Nations, the World Bank, the Association of Southeast Asian Nations (ASEAN), or the World Health Organisation. There are also international **non**-governmental organisations, such as Amnesty International, Greenpeace, or the International Federation of Red Cross and Red Crescent Societies.

At the local end, there are numerous not-for-profit groups likely working in your neighbourhood, such as:

- sports and recreational clubs
- community service organisations
- professional and business groups
- cultural societies
- unions
- co-operatives – member-owned business structures often used to strengthen local farming communities
- education – schools, of course, but there are many groups that focus on niche subject areas or that provide specialty educational tools, software, camps, or programs
- media – the demand for science communicators in journalism and reporting is on the rise.

Activity 3.8: Investigate How Organisations Are Structured

Activity goal: Explore the structure of a company or organisation.

Purpose and benefit: Develop awareness of jobs and management structures.

Activity steps:

1. Think of any job you've had, or one that you are familiar with, such as your parents' or relatives' workplace, or favourite fast-food outlet. It can be any organisation – a part time job, vacation placements, society/club roles, voluntary work, charity work, political or volunteer groups like sports or community activities (royal rangers). Pretty much anything that is organised by a group with a logo.

2. How big is the organisation? How many people work there?

   How is the company structured? This may be easier to draw. Start with yourself, then add lines to your boss, their boss, their boss' boss. Think about the people you work with and where they fit. Remember, this is just an exercise, there is no correct answer.

3. What does the business do? Does it sell a product? Provide a service?

   Who deals directly with the customer? Is there a sales team or front-of-house? How many people work with customers?

   How many people don't work directly with customers? This could include the boss, the finance or research team.

   What do you think these people do to support the company?

   Repeat this activity for other companies that you may be interested in. It is a useful exercise to understand how your skills can fit.

4. Repeat for 3 other companies that you might be interested in.

Takeaway: This is a useful exercise to understand how your skills can fit into an organisation.

Activity 3.9: How Does the Structure of an Organisation Impact the Workers?

Activity goal: Assess the structure of a company or organisation and identify the impacts on employees.

Purpose and benefit: Develop awareness of company culture and outline your own career priorities.

Activity 3.9: How Does the Structure of an Organisation Impact the Workers? – (*Continued*)

Activity steps:

1. Select a brand (*e.g.*, a fashion name) and repeat steps 2–6 with 3 different brands.

2. Research the company that owns the brand.

3. Identify the size of the company. Approximately how many employees does it have? (Mark on a sliding scale from small to big.)

4. Review their Careers page or About Us sections on their website.

5. How does the size of the company impact their employees?

6. Given your current priorities, is there a structure you prefer? *E.g.*, a smaller start-up may suit someone who prioritises the chance to try new things.

Scenario 3.3   Different Employers

Adrian started working for her family's restaurant when she was in high school. She was a quick and funny waitress, always remembering what her regular customers liked to order. Soon her parents gave her more responsibility in managing the till and counting the day's takings. After a few years she felt like she knew everything about running a restaurant. But she didn't see any room to grow or learn anything new. Even her parents encouraged her to try something different. They did not want her to do the same thing as them.

After graduating with a major in software engineering, she joined a big bank as an analyst, and soon realised that there was a specialty group that worked only on small business loans so businesses could renovate or build a bigger restaurant. This group of specialists came from a variety of backgrounds – accounting, management, even some with a psychology background. Joining this group, she felt like she was making a real difference in helping families like hers. While she didn't do much direct engineering, she used her data skills to identify key risks for her clients and calculated how much they could afford to borrow.

However, in her mid-30s, Adrian wanted to start a family of her own. Big banking culture was fun at times, with many perks like company dinners and social events. But it was also stressful, with long hours and an expectation

Continued

> **Scenario 3.3   Different Employers – (*Continued*)**
>
> of being well-dressed every day. Adrian knew it would be hard to manage such hours around school-times.
>
> With her experience, she quickly found a job working for a state government agency. The office was just around the corner and with shorter hours. She still applied her skills in managing infrastructure projects, where she was required to estimate costs and source equipment. After a few years, she was promoted to a senior management position. When she listed this on LinkedIn, she started to receive even more offers for similar roles at other government agencies.
>
> While Adrian's career was indirect, her STEM training and commercial awareness helped her identify career opportunities, even if they seemed a bit different.

"[My company] has a mentorship program. So I've been doing that. I have a mentor who is in the data team at the moment, even though I'm in the member services crew. I'm speaking to him and getting to learn a lot more about what it looks like in his position. It's super valuable in terms of, again, from a networking perspective, but also from just learning about the organization, learning about what their needs are, where I could fit into the team and give me a sense of where I might be able to progress if I wanted to move down into that area."

– Dylan, Bachelor of Arts/Science, major in chemistry and psychology, works in members services for a retirement finance company.

Products and Project Pipelines

As a STEM student, you have likely completed many projects throughout your studies. You are probably pretty familiar with the process. Start with an idea (or assignment), make a plan, assign tasks to team members, collate information, analyse the results and present the key findings by a certain deadline. Commercial projects can be similar, only much, much bigger. These could involve:

- research and development of a new product
- construction of a building or processing facility
- designing a new piece of equipment.

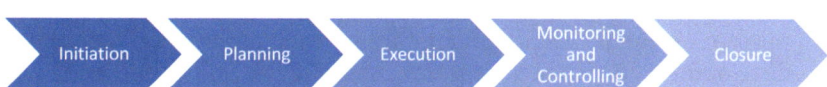

Figure 3.6   A project is a sequence of tasks with a clear beginning and end that must be completed to reach an objective.

While we have discussed how you can fit into a company or business structure, it is important to also recognise how you fit along a project pipeline. If you understand how projects work, it will help you to identify the type of work or 'stages' that you are most interested in yourself. You can move with a single project through different 'stages' (project management) or perhaps you prefer to focus on a particular stage that you can master over numerous types of projects. It may take some time (possibly years or decades) to figure out where your skills best fit along a project pipeline. However, understanding the general process is a good starting point to becoming more aware. Figure 3.6 summarises the stages of a project.

Projects are useful for STEM fields as you can break key problems into task-sized chunks that can be tackled by a number of people. In technology research, a project may be used to come up with a new design. For an already established product, a project may investigate a way to reduce packaging. Another project may look at improving employee safety. There is usually a benefit to the business for completing a project, either to improve a process or even to identify opportunities for improvement. Over the past few decades, project management and project systems have become established as a useful way to organise groups of people around a specific goal. Project Management itself is another STEM-related career path that many overlook.

Commercial Products and Services

Commercial products have the goal of generating profits. A company may sell a range of products or services. It is useful to consider the products that a company offers. This will help a STEM employee to recognise how their job supports the company's specific sales or services.

The Project Stages of a Commercial Product

Commercial projects can vary in size and scope and can involve a wide range of industries and sectors. However, these projects typically involve multiple stages. The following is an overview of the general stages of a commercial project:

1. **Research and development:** The first stage of a project involves conducting research to identify a need or opportunity. This may involve

market research, competitor analysis, and product or technology development. This stage requires many STEM skills and could start from pure research at a university through to clinical trials among the public.

2. **Business case development and feasibility studies:** Funding for any project or purpose is competitive, even within a single organisation or company, or even for a scientific grant. You need to justify any costs and explain why your case should be a priority. For example, the marketing team may want to develop a new line of skin care based on their customer surveys, while an operations manager wants a new centrifuge to purify liquids within hours rather than days. Both may be beneficial for the company, but clear business cases are needed for authorities to decide which project should receive funding. You may not need to make such decisions at this early stage in your career but understanding the process itself will help you to present your case when requesting funding. This will usually involve defining the objectives and goals of the project clearly, identifying what resources are required, developing a timeline, and estimating the costs and potential benefits. Financial terms are discussed further in the next section.

3. **Funding:** Once the planning stages are complete, funds are usually required to move forward, especially if equipment or large purchases are required. This funding can come from a variety of sources, such as venture capitalists, angel investors, crowdfunding campaigns, or bank loans. Costs and spending are typically tracked by a Project Manager.

4. **Legal:** A company must also ensure that it complies with all legal requirements before launching a product or service. Clarifying intellectual property, ownership, and how profits will be shared is critical at this stage. This also includes obtaining necessary licenses and permits.

5. **Execution:** This may involve starting construction, purchasing new equipment, or creating and launching a new product or service. This phase may include design and commissioning, where a new system or facility is tested.

6. **Close out:** A closeout stage is a critical part of any project. This identifies whether a project has been successfully completed and ensures that all loose ends are tied up. This could include settling outstanding invoices and ensuring that all parties are paid. The closeout stage also provides an opportunity to document the lessons learned, like identifying what worked well, what didn't work well, and any areas for improvement for next time.

Projects vary significantly across industries and around the world. These stages are only a guide to frame how your skills and experience can best fit into this process. For example, are you more interested in early-stage research, or would you rather see your product being made?

Project Manager

A critical role for any project is that of the Project Manager. Throughout each stage of a commercial project, it is important to track progress and manage people and spending. This will help ensure the project is successful and meets its objectives. Project managers are responsible for this planning, organisation, and execution. Their role is to ensure that projects are completed within the specified time frame, budget, and scope, while meeting the quality standards and stakeholders' expectations. While this role may not directly involve technical or scientific work, this position could be well-suited for graduates with STEM training.

Stakeholders

Stakeholders are a common term used in project management. This is another commercial structure that differs from the education sector. In school, there are teachers and students with clear roles of authority. Teachers give instructions. Students follow instructions. However, when you take on a job, you are responsible for completing a task. Many new graduates lack this ability to finalise a work assignment on their own. A common complaint by employers is that inexperienced workers lack confidence. It will be useful for you to recognise which stakeholders will help support your work.

You are expected to take ownership, but you are not expected to be left to flounder alone! Commercial stakeholders are other people or organisations that have a vested interest in your task or project. While they do not play the formal role of teachers that give you step-by-step instructions, these are people that share responsibility for decision-making and supporting the project. Your direct supervisor is typically your first go-to for assistance but other people in your company may also be responsible for supporting you and your work. There may be HR (Human Resources) Representatives that will organise training programs for you to learn. Another senior executive may have niche expertise in your project so your supervisor will recommend you speak to them. A vendor or salesperson may be able to offer advice on costs and delivery times that will help you plan a schedule. A local government agent may help you find a location to hold an event as they are responsible for developing cultural activities for the city. There are many stakeholders, formal and informal, that may share your mission and can help you in many ways. Stakeholders are people, groups or organisations with a vested interest in the decision-making and activities of a shared mission.

Formal stakeholders are usually organisations that are financially invested in a project. This is common in research and academic institutions where funding is provided by various sources – including government bodies, community groups, and industry associations. Working in these fields will often involve regularly communicating with members of these groups to share progress, understand their motivations, and work together towards a common goal.

A good start to any project is for you to first consider the stakeholders seen in Figure 3.7. Do you just need to do what your direct supervisor tells you? Are there several mangers giving you different opinions? Is the product for an external client? Who else has invested in this project? Are there legal or zoning regulations that need permission from government or environmental protection agencies? Do you need technical experts to give advice? Framing your work around how to best support and inform your stakeholders is mutually beneficial for all.

Figure 3.7 Company decisions are influenced by many stakeholders, both internal (orange) and external (blue).

Activity 3.10: Who Are the Stakeholders?

Activity goal: Identify and understand stakeholders and the part they play in projects.

Purpose and benefit: Build understanding of projects and stakeholder engagement. Supporting activity for early career roles and employees.

Activity steps:

1. When you have started a job (this can be a casual role while you are studying or a post-graduation role), write down one of the major tasks assigned to you.

2. Answer the following:
   - Who started the project?
   - Who assigned the project to you?
   - Who are the decision makers for your task?
   - Who is funding the work?
   - Is there anyone outside your discipline or area of expertise that can add useful or specialist input?

Takeaway: Understanding the stakeholders for a given project or task is extremely important. Understanding the stakeholders can reveal additional sources of information or help, can help you understand someone's concern about or interest in the project, and can help you understand how you should interact with these people. Telling the difference between the intrusive co-worker who wants to know what everyone is doing and the concerned specialists looking to mentor or help is very important. With stakeholder analysis you will be able to understand the dynamics of your workplace better.

Financial Awareness

Money matters. In today's world, financial literacy is a useful skill. Not only can it help you stay on top of your personal spending, but learning to discuss financial matters can help you communicate across multiple disciples outside of STEM.

Financial awareness will help you gain a greater perspective on your particular field or speciality. Every industry is influenced by certain economic drivers. Why do some things get cheaper and others more expensive? Why do people care about price of crude oil? What motivates not-for-profit organisations? How do you estimate costs?

But why do you need to care about finance if you are a STEM graduate? Well, an obvious priority is to make sure you get paid properly. Other benefits to financial awareness include being able to understand the costs and benefits of a project, and to be able to communicate with other departments and disciplines.

Project Finance

Have you ever applied for a scholarship or research internship at university? Have you ever asked your lecturers how their research gets funded? If you are a student maybe ask what they're working on outside of teaching. Most professors love talking about their research projects!

Research projects and students are often funded by a range of sources. This can be directly from the university, although many departments and disciplines have their own separate budgets. Governments may provide grants or tax incentives. Private investors may be looking for a competitive edge by advancing early-stage research and technologies.

Let's look at an example of a researcher investigating soil erosion. The Research Budget in Table 3.2 outlines a proposal they use to apply for a government grant. They also receive a sustainable community award

Table 3.2   Budget summary for research project.

Impact of Groundcover Management Technique on Avocado Trees Research Budget	
Funding	
Government Grant	$250 000
Sustainable Community Award	$40 000
Avocadoes Incorporated	$300 000
Total Funds	**$590 000**
Staff Expenses	
Administration	$20 000
Research Internship	$40 000
Postdoctoral Fellow 2 years full-time	$160 000
Analysis Technician	$60 000
Undergraduate Students 2 years part-time	$120 000
Total	**$400 000**
Other Research Expenses	
Field License Fees	$25 000
Travel	$40 000
Materials and Supplies	$35 000
Laboratory Analysis Equipment	$75 000
Conferences	$15 000
Total	**$190 000**
Total Project Costs	**$590 000**

from a local environmental group that has requested the funds be used to hire a research intern. Another source of funding is provided by an international avocado company. Avocado trees suffer from soil erosion and the company wants to better understand how to protect their crops and yields. These are examples of stakeholders (as discussed previously) that may be regularly involved with the project, providing guidance and support. A government representative may require monthly progress reports to understand how the research is tracking. The company may only want to see the final report and a presentation at the end to summarise the key findings. Some companies may be more involved and encourage regular meetings to provide on-going assistance. Table 3.2 is an example of a budget summary.

This is a simplified example, but the concept is consistent for any size of project. There are often several income streams and expenses to consider. As a researcher, you may not ever be asked to work directly on the accounting itself, but it is likely you will be asked at least for estimates of equipment or material costs to perform your work.

The Story of Money, Money, Money...

Money has a long and complex history that spans thousands of years. In ancient times, many items were used as a form of money. These items were typically scarce, durable, and easily recognisable, making them ideal for use as currency. These included shells, beads (made from stone, bone or glass), salt (the origin of the English word for "salary"), and even livestock.

The oldest coins ever found were believed to have originated around the Mediterranean Sea and were used between 1000 and 500 BC (an example is shown in Figure 3.8). The creation of coins required a

Figure 3.8 Electrum coin from Ephesus, 650–625 BC. Early coins were made of gold or silver alloys and stamped with images of gods and emperors. Classical Numismatic Group, Inc. Reproduced from ref. 14, https://commons.wikimedia.org/wiki/File:Triti,_Phanes,_625-600_BC,_Ionia_-_301224.jpg, under the terms of the CC BY-SA 3.0 license, https://creativecommons.org/licenses/by-sa/3.0/.

deep understanding of metallurgy, forces, the tools used, and even the accounting and bookkeeping systems needed to track and tally items.

As civilisation developed, the heavy weight of coins made it challenging to trade over further distances. The first paper currency was developed during the Tang Dynasty (800 AD) in China, as a form of a promise note (an official I-OWE-YOU). Paper money is much easier to transport than coins or shells or cows (Figure 3.9), and quickly became the primary form of currency throughout the world.

Over the past 1000 years, the use of debt and accounting became widespread. In the mid-20th century, credit cards were first introduced in America, revolutionising the way we pay for goods and services. Digital payments began to emerge in the early 1990s, allowing for instant transactions across the globe.

As technology has advanced, so too has the world of finance. Today, project management and project finance have become essential tools for managing large-scale projects. The rise of e-commerce has also had a significant impact on the commercial world. The first e-commerce transaction was considered to be for an online sale of a 1994 Sting album. Since then, convenient online transactions have exploded in popularity, allowing businesses to reach customers around the world.

The story of money is one of innovation, science, and technology. From bags of salt to the rise of Blockchain, the world of money continues to evolve and adapt.

Figure 3.9   Paper money is a lot easier to carry than coins and more detail can be included. Photo from Jun Rong Loo on Unsplash.

Project Accounting Thinking and Terminology

While we STEM graduates may love science for the pure sake of curiosity, in the modern commercial world, research is typically seen as an investment. Business language is useful for communicating the value of research to the wider population. Using standard terms also helps when resources or funding are limited. Table 3.3 contains three example projects. If Project A costs $2 million and Project B costs $1.5 million, just that information would lead you to pick Project B. However, Project B may cost more in ongoing maintenance costs, so over the long-term Project A will save more money.

Understanding your stakeholders and their motivations will also help you to frame grant applications and funding requests around what your investors are looking for. If your main client is an avocado company, you may suggest they select the project that focuses on increasing their crop yield. If funding is supplied by the government, they may be more interested in hearing how your work will lead to more jobs in the region. If you are working directly for an avocado pit-removing facility, your priority may be to install a new process that will reduce your maintenance costs with minimal capital expense.

Consider the details for 3 different projects, given in Table 3.3. The table illustrates the various factors to consider for a 'successful' project. You can see how hard it is to compare apples with dragon fruit. It is important to remember that while accounting seems purely numbers, financial decisions are made from a myriad of considerations (including conscious or unconscious biases). A CEO in charge of a publicly listed energy company may not want to invest a significant amount of cash on a new technology but may instead prefer to cut operating costs. A director of a small start-up may be happy to take the risk of funding research to develop a new product, but only if it can make a lot of profit in return. Some organisations have limits on their payback periods, such as banks that usually only provide house mortgage loans up to 40 years.

Financial metrics are just one way to describe the value of a project. In the next section, we will discuss how costs and benefits can be framed for a project.

Costs and Benefits

In daily life, we perform cost/benefits analyses all the time. Have you ever been running late and weighed up your travel options? Sure, a taxi would be faster but will cost a lot more than the bus. Is it worth it to prepare lunch from home or do you value the convenience of grabbing hot takeaway near the office?

> *"A cost/benefits analysis is a systematic method to assess various issues for a proposed plan of action."*

> – Rebecca Yee. Chemical engineer, former researcher, former business owner.

Table 3.3 Factors for a successful project.

	Project A	Project B	Project C
Project description	A community centre receives funding from a local media business to install a music recording studio for high school students.	A robotics company is developing an exoskeleton prototype that can help construction workers by supporting their back and providing bonus strength.	The government's Department of Transport is concerned about the drop in train users after COVID-19. They want to research which stations and lines are most impacted so they can better plan for investment in these zones.
Project lifespan or period	18 months	15 years	3 years
Expenses	Music recording equipment, building renovations, labour, marketing	R&D staff and equipment, legal and business administration	Cameras and monitoring equipment or software, staff, project management
Capital expense	$50 000	$3.5 million invested over 10 years	$120 000
Operating expense	$8000 p.a. for maintenance and security	$1.6 million p.a.	$250 000 p.a.
Expected revenue	None	The global exoskeleton market size is expected to grow from US$ 0.7 billion in 2023 to US$ 3.7 billion by 2028.[2]	None but this project will provide useful background information for future investments and projects.
Payback period or return on investment	N/A	Will depend on success of commercial sales.	Unknown
Non-quantifiable metrics (social, environmental, etc.)[a]			

[a]For you to fill out!

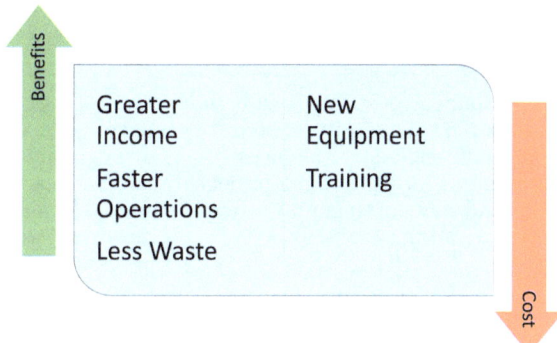

Figure 3.10 Most of our decisions come down to weighing up the costs and benefits.

In a commercial project, you are also limited by the same factors. Time and money. When you first start working, it's unlikely that you will be making major project decisions. However, you will possibly be required to collate information for your supervisor. Many jobs now require more than just such a summary of information. You will often be asked for your recommendations and reasons for such recommendations. You may be asked to put together a business case or argument for one solution or another.

A cost/benefits analysis (represented in Table 3.4 and Figure 3.10) is a systematic method to assess various issues for a proposed plan of action. It is important to remember that, as much as we STEM workers love numbers, decisions are often made from the heart (many millions of dollars in funding have been directed to the old schools or colleges of wealthy individuals). While these factors can provide insight and perspective, individuals can also push action in certain directions based on their primary concerns and opinions about what is most important. Different people and groups will have different priorities. Table 3.4 lists several common concerns for many STEM and general projects. These are useful to consider, but many other issues can be taken into consideration.

Legal Awareness

This is a reminder that you do not need to be an expert in every field. However, a basic appreciation of contracts and legal obligations will be useful even for STEM graduates, especially when you start working. You may already be familiar with your legal rights and responsibilities as a student. When you enrolled at college or university, you likely signed an Enrolment contract or a formal acceptance letter. These sometimes include details of your obligations, such as class attendance, socially appropriate behaviour, not bringing drugs onto campus, proper clothing, *etc.*

Table 3.4 Cost/benefits analysis.

	Costs	Benefits
Financial	Financial costs are usually the easiest to quantify. Expenses could include initial investments for equipment, project and staff costs, maintenance, operating expenses.	The potential financial benefits are usually seen through increased revenue, reduced operating costs (savings over time), improved productivity or process efficiencies.
Risk	Risk is more difficult to quantify, although systematic assessments are often used to highlight the 'biggest' risks through a likeliness-*vs.*-consequences Risk Matrix. These could include increased hazards or reduced safety measures, reputational damage, and legal liabilities.	Many projects aim to reduce risk by implementing something that will result in safer or smoother operation. This may remove hazards or put improved safety measures in place, such as installing a handrail. Safer operation is a competitive advantage to attract the best staff.
Operational	In addition to financial expenses for operational costs like staffing and maintenance, operational costs can also include increased complexity or reduced safety. For example, introducing a new gas turbine.	Operational benefits can include improvements to the process, such as increased efficiency or productivity, more streamlined work systems, or safer working environments.
Social	Negative social impacts could include reduced employment (from the closure of a facility), reduced places to socialise (such as a carpark being built on a park), negative health impacts on communities or individuals (from pollution).	Social benefits could be an improvement to the community's quality of life, more jobs and economic opportunities, as well as safer and cleaner environments.
Environmental	Our understanding of the broader costs and impacts of ecological damage is still developing. Pollution and reduced biodiversity are noticeable costs that are challenging to quantify, especially in financial terms. Negative environmental impacts could include reducing population numbers or diminished quality of natural resources like water.	Environmental benefits will often tie into improved societies as well. Healthier biodiversity and ecosystems provide wider benefits to all. Benefits could include reduced pollution or increased quality of shared natural resources, such as air.

For STEM jobs, legal structures such as Partnerships and Joint Ventures are common, especially for research projects with multiple stakeholders. Funding may only be released on the dates specified in these agreements. It's useful to be aware of how organisations cooperate and use contracts to establish boundaries. These boundaries outline who is responsible for what, as well as who will control or 'hold' the benefits of the work once it's done.

Setting up such legal structures and documents can be time-consuming. Contracts may need to be reviewed or signed by several people across different organisations. Stakeholders may want certain legal terms or clauses in place to protect certain trade secrets. In many cases, research is not legally protected until such documents are signed. It is important to recognise that legal processes take time and must be set up properly. Taking the time and attention to work out legal boundaries and ownership can help avoid disagreements and more legal expenses later.

How to Understand Job Contracts

An Introduction to Employment Contracts

Your first formal introduction to the commercial world will likely be an employment contract for a job. These are contractual arrangements that outline the terms of your work and compensation. An example can be found in Figure 3.11, with details listed in Table 3.5.

This is a general employment contract. Some countries and specific jobs may have variations or additional clauses, but these key features should be a part of any reasonable agreement.

Employee Rights and Responsibilities

It is highly recommended that you ask someone else to check your employment contract. Especially if you are new to the workforce, a more experienced eye or mentor may draw your attention to something you wouldn't even notice, such as no mention of sick days or mental health leave if that is usually standard.

Employee rights vary widely internationally, depending on the country's legal system, political climate, and cultural norms. Keep in mind that many corporations will have variations or additions to these minimal rights, such as offering more paid leave for holidays to attract the best workers. Here are several factors to consider as an employee:

- **Legal working hours** – Is there are a maximum number of working hours per week? Do you have compulsory break time or rest periods between scheduled shifts?

<Company letterhead logo>

<Date of the Letter>

Private and confidential

<Your full name>
<Your residential address>

Dear <Your name>

Letter of engagement

I am pleased to offer you employment in the position of < position title> with us at <trading name of business> ('the employer') on the terms and conditions set out in this letter.

1. **Position**

   1.1 Your start date will be < start date>.

   1.2 Your employment will be <full-time/part-time>.

   1.3 The duties of this position are set out in the **attached** Position Description. You will be required to perform these duties, and any other duties the employer may assign to you, having regard to your skills, training and experience.

   1.4 You will be required to perform your duties at <location>, or elsewhere as reasonably directed by the employer.

2. **Probation**

   2.1 A probation period will apply for the first <THREE> months of your employment. During this time we will assess your progress and performance in the position.

   2.2 During the probation period you or the employer may end your employment by providing notice in accordance with the table in clause 8.1 below.

3. **Terms and conditions of employment**

   3.1 Unless more generous provisions are provided in this letter or in the attached Schedule, the terms and conditions of your employment will be those set out in the <relevant award or enterprise agreement name> and applicable legislation. This includes, but is not limited to, the National Employment Standards in the *Fair Work Act 2009*.

4. **Ordinary hours of work**

   4.1 Your ordinary hours of work will be <number of hours: 38 if full-time> per week, plus any reasonable additional hours that are necessary to fulfil your duties or as otherwise required by the employer.

5. **Remuneration**

   5.1 You will be paid <weekly/fortnightly/monthly> at the rate of $<XX> per <hour/week/month/year>.

   5.2 The employer will also make superannuation payments on your behalf in accordance with the *Superannuation Guarantee (Administration) Act 1992*.

   5.3 Your remuneration will be reviewed annually and may be increased at the employer's discretion.

6. **Leave**

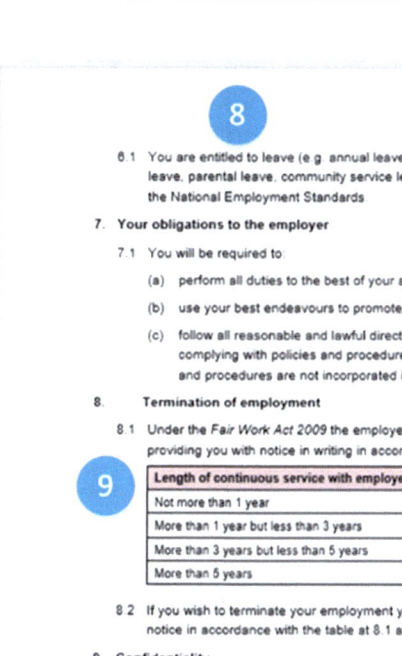

6.1 You are entitled to leave (e.g. annual leave, personal leave, carers leave, compassionate leave, parental leave, community service leave and long service leave) in accordance with the National Employment Standards

7. **Your obligations to the employer**

   7.1 You will be required to:

   (a) perform all duties to the best of your ability at all times;

   (b) use your best endeavours to promote and protect the interests of the employer; and

   (c) follow all reasonable and lawful directions given to you by the employer, including complying with policies and procedures as amended from time to time. These policies and procedures are not incorporated into your contract of employment.

8. **Termination of employment**

   8.1 Under the *Fair Work Act 2009* the employer may terminate your employment at any time by providing you with notice in writing in accordance with this table:

Length of continuous service with employer	Period of notice
Not more than 1 year	1 week
More than 1 year but less than 3 years	2 weeks
More than 3 years but less than 5 years	3 weeks
More than 5 years	4 weeks

   8.2 If you wish to terminate your employment you are required to provide the employer with prior notice in accordance with the table at 8.1 above.

9. **Confidentiality**

   9.1 By accepting this letter of offer, you acknowledge and agree that you will not, during the course of your employment or thereafter, except with the consent of the employer, as required by law or in the performance of your duties, use or disclose confidential information relating to the business of the employer, including but not limited to client lists, trade secrets, client details and pricing structures.

If you have any questions about the terms and conditions of employment, please do not hesitate to contact HR or Manager's Name. To accept this offer of employment please return a signed and dated copy of this letter to me by < date>.

I, <insert name of employee>, have read and understood this letter and accept the offer of employment from <insert company/partnership/sole trader name> on the terms and conditions set out in the letter.

Signed _____   Date _____ / _____ / _____

Print name: _____

PLEASE KEEP A COPY OF THIS LETTER FOR YOUR RECORDS

Figure 3.11   An example of an employment contract.

Table 3.5 An example of the basic format for an employment contract.

1   **Company letterhead**
     Familiarising yourself with the company website and history will serve you
     well in your job search. Try to understand the company's culture to see if
     you will fit well with the people. Check that the company logo and details
     are consistent to ensure you are dealing with a reputable organisation.

2   **Position title**
     Your job title will usually represent whether you are early, mid-, or
     late in your career. Look up the definition of new positions to help you
     better understand career options and pathways.

3   **Full-time/Part-time & start date**
     It should be clearly stated what day you commence working for payroll
     (make sure you are paid for the hours/days you have worked) and legal
     reasons. Your expected commitment should also be shown, with your
     expected hours if casual or part-time.
     Full-time: ~40 hours per week,[a] usually around business hours (9 am–5 pm)
     Part-time: A set number of hours or days with partial benefits (such as
     long-service leave)
     Casual: Most flexible for both employer and employee – an indefinite
     agreement with no commitments for future work. Working hours
     are essentially agreed upon as needed by both parties, usually with a
     schedule set in advance *e.g.* your shifts for the week will be provided a
     week prior. Casual hourly rates are usually higher to compensate for job
     uncertainty and a lack of benefits such as sick leave.

4   **Position description**
     You may have already seen this as a separate attachment or used in
     advertising the job. A clear job description should outline your role and
     responsibilities.

5   **Probation**
     A trial period may/will be required for many jobs. This may involve
     reduced pay or hours for a set time. Employers use the probationary
     period to assess whether the new hire is a good fit for the position.

6   **Award or enterprise agreement**
     Minimum wage
     Agreements like superannuation

7   **Remuneration**
     Pay and other benefits (such as retirement fund contributions)

8   **Leave**
     Most countries will have a set amount of leave

9   **Termination of employment**
     Process and dispute resolution

Table 3.5 *(Continued)*

10 **Confidentiality and IP**

Protection of IP
Depending on your industry and the type of work the person will be doing, you may wish to seek independent legal advice about the protection of intellectual property.

11 **Contact details**

If you are uncertain about anything in the contract, you should feel confident to contact the company's HR manager or your new supervisor with any questions. A company should not pressure you to sign an agreement until you are comfortable with the terms.

ªNote what is considered full time in different parts of the world can vary.

- **Paid leave** – Most countries have laws mandating paid vacation time, sick leave, and parental leave for employees. However, the amount of leave and the conditions vary widely.
- **Minimum wage** – Some countries have laws mandating a minimum wage that employers must pay to their employees. However, the level of the minimum wage and the criteria for eligibility can vary depending on a nation's economy.
- **Health and safety** – Does your country have laws mandating that employers provide a safe and healthy workplace for their employees? Does this include mental health and discrimination issues? Many countries have laws prohibiting discrimination based on factors such as race, gender, and sexual orientation.
- **Confidentiality** – While these requirements are more likely to be specified at an organisational level, this is still another important consideration for employees. Check your contract for any legal conditions on protecting information. Confidentiality is a critical aspect of protecting intellectual property in STEM fields. STEM workers often deal with proprietary technologies and trade secrets that need to be safeguarded. To ensure that sensitive information is only accessed by authorised personnel, employees may be requested to agree to certain terms or agreements, such as Non-Disclosure Agreements (NDAs). These are signed legal contracts that essentially promise that you will not disclose certain information to outside parties. NDAs can be used to protect proprietary information between organisations that may be collaborating on a shared project.

It is important for workers to be aware of their rights. Many countries have labour unions, legal aid clinics, and even online resources that can help workers understand and defend their rights in the workplace.

Activity 3.11: Identify Your Relevant Employment Rights, Governing Bodies, and Agencies

Activity goal: Find the agency that oversees your employment rights. Recognise your rights as an employee.

Purpose and benefit: Build knowledge of employment rights.

Activity steps:

Look up your local employment rights.

1. Look up your employment rights and answer the following questions:
    - Are they listed on a reliable government website?
    - Is there a department or agency that enforces regulations with fines or shut downs?
    - Are there any non-government organisations that also provide guidance around employment rules?

2. Identify any of the following relevant information on rights in your area:
    - legal working hours
    - paid leave
    - minimum wage
    - health and safety responsibilities of the employer.

3. If you are looking to work in multiple places, repeat with at least one more location.

Takeaways: Each jurisdiction can vary, even within the same country. Although state to state variations are not usually major, it's important to investigate. By knowing what your rights are you will be able to protect yourself in the future, but you will also be able to protect yourself by choosing a job that has better conditions. Alternatively, you will know if questionable companies are trying to tell you something is legal for them to put in your contract when maybe it isn't.

Intellectual Property

> "Intellectual property (IP) refers to creations of the mind, such as inventions; literary and artistic works; designs; and symbols, names and images used in commerce."
>
> – World Intellectual Property Organisation.[3]

Barbie® dolls are one of the world's most recognisable toys. First produced by Mattel in 1959 (over 60 years ago), Barbie® was the most popular doll in America for decades, until Bratz® were launched in 2001.[4]

Bratz® dolls quickly took over 40% of the market share for fashion dolls. In response, Barbie® released a new line of younger style dolls with big heads, slim waists and wide hips – a very similar design to Bratz®.

In 2008, MGA (the manufacturers of Bratz®) filed a claim against Mattel claiming design copyright. Mattel counter-sued by arguing that the designer of the Bratz® dolls was working for them when he created the design.[5] A long legal battle continued between the corporations. The jury ruled in favour of Mattel and ordered that MGA pay $100 million in damages and remove Bratz® dolls from sale for one year.[6] The IP case became famous as the issue was debated – can an individual's creative work be completely owned by an employer or do boundaries apply if the work is produced outside of contracted hours? This isn't just about toys. Bratz® dolls have made MGA US$ 1.9 billion and Barbie's maker, Mattel, has a net worth of $6.8 billion.[7]

For STEM employees, your work is commonly considered as Intellectual Property. This can include reports, spreadsheets with calculations, unique models, or even new technical processes. IP rights are usually defined in your employment contract. Part of this contract may also state that you are responsible for helping to protect your company's IP – by using passwords on devices and making sure to shred sensitive paperwork after the project is completed. If you are part of larger projects, IP ownership is often outlined in partnership agreements or contracts. This establishes decisions like whether Company A has the right to profit from the research or if the report must be openly published and shared to benefit the wider industry.

"Working in IP in the UK and Europe is very structured but it is also a lot more grey and less black and white than science training."

– Jonathan, physics/nanotechnology graduate and patent attorney, not working in science but still at the forefront of science.

Intellectual property rights are meant to stimulate innovation. Intellectual Property includes abstract products like inventions, trademarks, designs, creative content like music or other intangible assets. The IP system allows creators to protect their new ideas and benefit or profit while this protection is in place. An IP holder can profit not only from direct sales but also from licensing to third parties for a fee or royalty. This means that the inventor does not have to manufacture and commercialise the products themselves. This practice is common in STEM fields. An inventor may be a postgraduate student researching an early-stage technology, which is patented by the university. The patent states that the university 'owns' the technology, but the inventor is legally entitled to 20% of royalties. A private company licenses the patent from the university and mass manufactures a new product. The inventor will receive 20% of the what the company pays the university for the license but will not receive anything from product sales. Alternatively, the inventor may be hired directly by a company to design a new product. Such an employment contract might state that the inventor has no ownership of the product but would receive 5% of the profits from every product sold.

100 Years of $1 Insulin

Sir Frederick G. Banting is credited as the 'inventor' of insulin. Insulin is a hormone produced in the pancreas that regulates the amount of glucose in the blood. People with diabetes do not make enough insulin to keep their blood sugar at healthy levels.

Sir Banting was a surgeon, not a scientist. In 1920, he partnered with John Macleod and his research student, Charles Best, to conduct experiments at the University of Toronto. Insulin removed from a dog's pancreas was shown to reduce blood sugar levels. They were then joined by a biochemist, James Collip, who helped to purify the insulin to be safe for human testing.

Banting, Collip and Best were awarded the US patent for insulin and the method for making it. However, they sold these patents to the University of Toronto for $1. Banting famously said, "Insulin does not belong to me, it belongs to the world." Banting and Macleod were jointly awarded the 1923 Nobel Prize in Physiology or Medicine. Both split their prize money with Best and Collip.

Several pharmaceutical manufacturers have since licensed the rights to the patent and can mass produce commercial insulin using Banting's method. Over the last century, manufacturers have made new improved insulins through different methods. There are now numerous additional patents for insulins, as well as secondary patents on non-active ingredients or delivery devices like insulin pens.[8] Since the operational equipment for making insulin is quite complex and expensive, some manufactures don't even have a patent-protected product, they rely on their brand-name or other IP tools like trade secrets (operators sign a contract to confirm they will not disclose special equipment and processes to outsiders).

Understanding the various tools used to protect Intellectual Property will help you to structure agreements to maximise your benefit. As with many legal issues, planning is the key to avoiding problems in the future. Different methods of IP protection may be useful for different organisations. The cost and timing of putting different types of legal protections in place can also vary.

The 4 main types of Intellectual Property protection include:

- patents
- trademarks
- copyrights
- trade secrets.

Patents

When introduced to intellectual property, patents are usually the first thing that comes to mind. Patents are used to protect inventive ideas and processes. To file for a patent, descriptions, prototypes or drawings can provide proof of the inventive process and define what the invention actually is. The patent is reviewed by a patent office and awarded if the invention is considered 'new, useful and not obvious'.

The patenting process is complicated, expensive and often takes experienced patent attorneys and several years between application and award. Costs will vary depending on the country or countries where you file for protection, usually in the range of tens of thousands of dollars. Most individual inventors will work with universities or institutions to support this process in exchange for a share of ownership or profits.

Once a patent is approved, it has a set lifetime. Patent holders usually try to sell or license the invention to an established company that has the brand or manufacturing capabilities to bring the invention to market. One important recent patent you might not have thought of but may have made use of is the patent for plastic bank notes (Figure 3.12). Another famous but much older patent is that for rivet work pants aka denim jeans (Figure 3.13).

Figure 3.12 Polymer bank notes developed by CSIRO in Australia – US Patent Number 4536016.[9] Photo from PiggyBank on Unsplash.

Figure 3.13 Riveted denim work pants – US Patent Number 138121.[10] Photo from Tomas Martinez on Unsplash.

Trademarks

A trademark is unlike a patent in that it can protect words, phrases, symbols, sounds, smells and even colour schemes. Trademarks do not require such an extensive approval process compared to patents, usually just registration at a relevant government office is sufficient. These usually only cost a few hundred dollars, but the value is part of its marketing power. Trademarks are valued for branding and identity. A good trademark is instantly recognisable and associates a certain product or service with a specific organisation.

Copyrights

Copyrights protect the expression of ideas, usually in the creative space rather than a commercial product or process. Original works, including literature, art, music, architectural drawings, or even programming code, can be registered for copyright. This protection allows the owner to control production, performances, adaptations, and distributions. Copyright registration applications are much simpler and cheaper than patents or trademarks.

Ice Ice Baby

Vanilla Ice, aka Robert Van Winkle, released *Ice Ice Baby* in 1983. The song has a similar bass line of *Under Pressure*, released by Queen and David Bowie two years prior. Queen and Bowie threatened to sue for copyright infringement, but the matter was settled out of court, with Van Winkle paying an undisclosed financial sum as compensation.

Copyright Cultural Clash[11]

Andy Warhol is an American artist best known for his pop art style. Warhol's art is influential as it challenges the very notion of copyright. His painting method uses silkscreen printmaking, which Warhol used to repeat particular pop culture images, such as celebrities (like Marilyn Monroe) and common commercial products (like Campbell's soup can).

Warhol often used other artworks as a starting point. This was intentional to highlight the reproducibility of commercial popular culture. In

1984, Warhol sourced a photo of the musician Prince, and drew over it to create his signature multi-coloured series of prints for a cover of Vanity Fair magazine. The photographer, Lynn Goldsmith, received a license and credit for this publication only. However, Warhol produced several additional paintings, prints, and drawings in this series, using the same photograph. In 2016, in the wake of Prince's death, Vanity Fair published these new works without crediting the original photographer.

The Andy Warhol Foundation filed a pre-emptive lawsuit in 2017, asking the court to state that the Prince Series was *not* in violation of Goldsmith's copyright. Goldsmith then launched a countersuit just a few days later, claiming for years of unpaid royalties. A federal judge ruled in favour of the Warhol Foundation in 2019, saying the painting was significantly "transformed" because it conveys a different message from the original, and thus is fair use.[12]

However, a second federal appeals court later reversed the decision in 2021. This case has since been taken to the US Supreme Court with The Warhol Foundation arguing that Warhol's appropriation of Goldsmith's work counted as 'fair use'.

Unfortunately, Andy Warhol passed in 1987, so we cannot know what the artist himself would have thought. Many artists and media groups have contributed their opinions to this ongoing court case, including the American Association of publishers, the Motion Picture Association of America, the Digital Media Licensing Association, the Recording Industry Association of America, and the Screen Actors Guild-American Federation of Television and Radio Artists.

The Supreme Court's decision could provide more protections for original artists, but this outcome would also restrict artists from building on existing work. In these days, where digital content is so easy to replicate and memes are part of daily communication, copyright issues can be challenging.[13]

Trade Secrets

Trade secrets are simply that. Secrets that companies and organisations try to keep to give them a competitive advantage. Formal federal registration or legal paperwork is not required but confidentiality strategies can include:

- confidentiality contracts, such as Non-Disclosure Agreements (NDAs),
- proprietary procedures,
- secure IT and paperwork filing systems,
- password protected devices,
- data security measures, keeping logs of visitors and restricting access.

Companies need to take proactive actions to maintain secrecy. Employees will be notified of sensitive information and company practices to protect this content. Nondisclosure agreements are also commonly used. These are typically contracts used to make it clear to all parties that there is sensitive content to protect.

Coca-Cola® is famously known for keeping its recipe a trade secret rather than a patent. To obtain a patent, they would have had to disclose the recipe, which would then be available to the public after 20 years. Instead, they have kept the Coca-Cola® formula as a trade secret for over a century. According to urban legend, only two employees know of the complete formula at any given time and are not permitted to travel together.

But what happens if IP protections fail? Even with the best intentions and careful legal protections, parties could disagree about the results afterwards. One side may want more than the original agreement, as they feel they have contributed more than originally agreed. Unforeseen situations can create expensive legal battles. However, this is not the end of the world. Companies with sensitive IP usually have legal teams that can provide guidance through any litigation issues. Individuals usually have little direct risk from IP infringements. However, being involved in such cases can negatively affect your career or employment.

Much of your 'value' as a STEM graduate is your knowledge, understanding, analytical and creative mind. You may come up with a new method for taking samples, a new software code for compiling data faster, even a new heat-sensitive material that could revolutionise drug delivery – these inventions are Intellectual Property. Being aware of the legal requirements and tools ensures you are recognised and protected as the creator of an invention; literary and artistic works; designs and software.

Depending on your job contract or funding agreement, such IP may not solely be yours. And that's OK – the real value in IP is being able to use it. What do you really want from your invention? Is it essential that your name be listed as the inventor? Do you just want fair financial compensation? Is it important that you maintain control of the invention's development? What will happen to the IP structure if you leave that organisation?

How to Grow Your Commercial Knowledge

"When you are at university, you just don't know what opportunities are out there. So, expose yourself and then you can know what your interest is, what you're passionate about, and then it's easier to plan for your future and really narrow down what you want out of your career. And that opens a lot of doors for you."

– Ava, mechanical engineer, master's in biotechnology, PhD. Worked as an engineer for several years. Started her own business. Currently head of marketing and sales.

By being commercially aware, by the time you come to an important job interview or major career decision, you will have a broader understanding of the bigger picture and be able to choose a suitable direction for your own path.

"You don't know what you don't know."

You can grow your commercial awareness by subscribing to industry newsletters, attending networking events or even just following news in your field. Following commercial news can help you get familiar with the big players working in your field of interest. Who are the companies that get discussed frequently? Is it in a good light or are they being accused of problematic activities? There are also many government websites that provide useful outlines of funding structures and big-picture commercial opportunities.

Activity 3.12: How Can I Build My Own Resources to Learn About the Commercial World?

Activity goal: Sign up to industry publications.

Purpose and benefit: Take ownership of your own professional development by planning to regularly consume industry/business news you can learn from.

Activity steps:

1. Find 2–3 industry-specific or business news publications or websites to read. First list any industries that may suit your skills and interests.

2. Look up podcasts or news outlets that report on your industry and sign up to email alerts to be notified of new episodes or releases. As these come in some might not suit so unsubscribe and look for a better resource.

3. Repeat this to get a broad mix of sources that will balance any bias that the publication(s) might have and to get a wide array of news/information.

4. Schedule a regular habit of learning. Putting something in the diary for a regular time that doesn't get disrupted is a good approach. For example, put a podcast on during a commute.

Takeaways: Keep in mind that this activity is most useful the earlier you can do it. These resources will present a lot of new concepts you may otherwise never have learned about in your formal studies, but it will take time for you to digest new vocabulary and the new ways of thinking that you are likely hearing in some of these publications. This learning can also give you additional keywords to use when searching for specific jobs.

Employers like to see that you have thought about their business and their needs. This can set you apart from other graduates who could do the same job. When you go into the job interview, can you show that you understand the company business model? Are they owned by public shareholders or a few private founders? Do they care about employee satisfaction? Are they a government organisation or impacted by government regulations (such as defence security or utilities like water that must be a certain quality for public use)?

Visiting a company website is a good starting point. Look for any press releases (formal statements by the company) or news articles. Are they part of a broader industry? Are they publicly listed on the stock market? What are the trends of the industry? Do they go up and down, or is it a stable market? What type of issues or events would most impact the organisation?

Activity 3.13: What Can I Research About a Specific Company?

Activity goal: Practice investigating an organisation's commercial background. Repeat this task every 4 months.

Purpose and benefit: Learn how to develop your own commercial awareness.

Activity steps:

1. Select an organisation that you would like to work for. It could be a company, government organisation or non-profit charity. Write down this group's name. If you don't have one in mind, search online using a subject of interest (e.g., "water chemistry + company + graduate").

2. Look for commercial information on the company website.
   - Who owns the company?
   - How many employees are there?
   - Where is the head office and are there other sites or offices?
   - What do they make or sell?
   - Is their product or service one of the best?
   - Who else offers these products or services?
   - How is this company different?

3. Now explore if the company has been in the news by doing an online search for the company name and news.

4. Are there any news articles about them? Write down any major news about the company that you find.

Activity 3.13: What Can I Research About a Specific Company?
– (Continued)

5. Now imagine you applied to this organisation and have now been asked to come in for a job interview. You are confident with the technical questions. However, the HR (human resources) representative knows the company has a problem with younger employees leaving after a few years because they want to move to a bigger city or change industry. Managers wants to gauge if your interests are a good fit and aligned with their long-term strategy. You should be doing the same thing.

6. The interviewer asks, *"Is there anything you want to ask us?"*. Write down how you would respond to such a question. Think about your goals, or even just general desired lifestyle, and what questions would help you understand the company culture. For example:

- How big are the teams?
- Do people work in groups or mostly individually?
- Does your manager use online meetings or prefer in-person?
- Does the manager have a scientific background?
- Do any of the senior leadership team?
- Do you want a career with potential to move locations?
- Do you want security to know you don't have to travel if don't want to?
- Do you want a clearly defined professional development programme to achieve a specific management role or do you want the freedom to pursue a variety of job types?

7. If you have someone to practice with this is a good activity to bring in a helper who can help "interview" you or give advice. If you aren't ready for this activity yet, try steps 1–4 then wait until you have read Chapter 6: Resumes and Cover Letters and Chapter 7: Addressing Key Selection Criteria and Interviews, which will help you with preparing for applications and interviews, and then return to complete this activity.

Takeaway: Commercial awareness can feed back into both the job search process and the actual interview. Plus, of course, your decision whether or not a company is right for you. As you build your commercial awareness you will discover more and more examples of how this is true.

These are complex issues, and it takes time for a student, new graduate, or even an experienced worker to really grasp how entire industries operate and their place within it. So don't leave these exercises to the last minute. It's not exactly something you can memorise and cram the night before. Start exposing yourself to commercial conversations early. This chapter touches on commercial concepts such as industry sectors, economic structures, financial and legal issues. Feel free to read through or flick to the sections you want to know more about. And share or discuss with your friends and family – you may discover that someone has a wealth of unexpected knowledge in a field of your interest.

Commercial awareness is essential in a rapidly changing working world. This will help you to identify future trends, new job opportunities, and workforce disruptions that may impact your career prospects. For example, the COVID-19 pandemic has had a significant impact on the workforce and how we think about globalisation. Office jobs have been revolutionised with changes such as work-from-home technology and international mobility, while on-site or field jobs have become even more in demand. Being aware of these disruptions will help you understand how the job market may shift in the future.

Emerging fields such as advanced manufacturing, new energy, digitisation, and the circular economy offer exciting opportunities for those with STEM skills and knowledge. Learning about these new industries and the type of organisations that work in this space can give individuals a competitive edge in the job market. Economic changes will also impact job opportunities. Commercial awareness is not just about knowing the current business landscape, but also about being aware of future trends and changes that can impact the workforce.

Who Else Can Help?

Developing commercial awareness is an important skill for anyone, regardless of their profession. Hopefully, this chapter has alerted you to what else is out there. It is essential to have a broad understanding of the commercial landscape, including finance, marketing, legal and regulatory issues, and other key areas. But you are not expected to be an expert in these fields! There are many experts and support services that can provide specific financial or legal advice and guidance. There are even services for career coaching, mentoring or even job seeking and resume writing advice. For personal finance, many banks will offer services where you can speak one-on-one to a financial or business advisor. There are many online resources and formal training programs around commercial topics that you may find interesting too!

By developing your commercial awareness, hopefully you will be better equipped to navigate a fulfilling and successful career. This awareness will help you identify opportunities and make informed decisions in many areas of your life.

Don't be afraid to ask people around you for their perspectives and insight. Seek out the resources and support you need. Really, just keep the conversation going and you never know what you will find out next.

Check Your Progress

In the first chapter you likely filled in a short survey to gauge your perception of your employability, which will have been emailed to you. This was called a self-perceived employability questionnaire (Activity 1.1). The subject on the email will have the title of the book in it. If you are ready to check your progress, use this QR code to recomplete the questionnaire  and it will send the new results. Compare these results with any you have been sent previously by doing the following:

1. Bring the results up together and compare your results. If you are later in the book, you might have more than two. You can compare which results you want, or even all of them. Transferring the data to an Excel spreadsheet will help if you have many answers and are inclined to do that, or if you need an activity to help develop your Excel skills.
2. What are the differences? Sometimes you will feel a bit different on a different day, so not every value that changes by one point is a significant change. Maybe you had a good day or a bad day. Can you see any trends or big differences? Do any of the changes feel particularly true? Even if it is just one point.
3. Often when people learn more about a subject, they first loose a bit of confidence as they are starting to learn about how much there is to learn. Has that happened to you? This is an important process to be aware of. Maybe you aren't feeling as comfortable with your employability but that's because you had to realise how much there was to know to be able to go out and learn it. Well done, that was an important realisation.
4. Now that you have stood back and thought about your overall perceptions of your employability, are there any plans you think you might make or change? Or maybe something you have realised you want to prioritise or deprioritise? If so, make a note of these and set a calendar reminder once a week for the next month to ensure it doesn't get forgotten.
5. Enjoy the next chapter!

References

1. Australian Bureau of Statistics, Employment and Earnings, Public Sector, Australia Report - 2021-22 financial year, 2022, https://www.abs.gov.au/statistics/labour/employment-and-unemployment/public-sector-employ-ment-and-earnings/2021-22, accessed on 18th February, 2024.

2. Transparency Research, Exoskeleton Market Size, Industry Report, Growth Drivers, Trends, Analysis, 2032, 2022, https://www.transparencymarket-research.com/exoskeleton-market.html, accessed on 10th January, 2023.

3. The World Intellectual Property Organisation, What is Intellectual Property? https://www.wipo.int/about-ip/en/, accessed on 10th February, 2024.

4. Edology, 4 Famous Intellectual Property Cases, 2022, https://www.edology.com/blog/law-criminology/4-famous-intellectual-property-cases/.

5. ABC News, Barbie Plays Dirty, Bratz's Dirty Tricks Suit Claims, 2014, https://abcnews.go.com/Business/barbie-plays-dirty-bratzs-dirty-tricks-suit-claims/story?id=21541339, accessed on 4th November, 2023.

6. M. Gambino, Ten Famous Intellectual Property Disputes, in *Smithsonian Magazine*, 2011, https://www.smithsonianmag.com/history/ten-famous-in-tellectual-property-disputes-18521880/, accessed on 9th October, 2023.

7. MacroTrends, Mattel Net Worth 2010-2022, 2024, https://www.macrotrends.net/stocks/charts/MAT/mattel/net-worth, accessed on 18th February, 2024.

8. J. Belluz, The absurdly high cost of insulin, explained in *Vox*, 2019, https://www.vox.com/2019/4/3/18293950/why-is-insulin-so-expensive, accessed on 18th February, 2024.

9. C. Ward, Polymer banknotes publications, in *CSIROpedia*, 2011, https://csiro-pedia.csiro.au/polymer-banknotes/, accessed on 17th December, 2023.

10. Levi Strauss and CO, The History of Denim, 2019, https://www.levistrauss.com/2019/07/04/the-history-of-denim/#:~:text=May%2020%2C%201873%20marked%20an,for%20the%20very%20first%20time, accessed on 1st August, 2023.

11. S. Song, Andy Warhol Copyright Infringement Case Will Go to the Supreme Court, in *Papermag*, 2022, https://www.papermag.com/andy-warhol-copyright-infringement-case#rebelltitem3, accessed on 25th October, 2023.

12. T. Waite, Andy Warhol loses copyright battle over Prince art from beyond the grave on *Dazed*, 2021, https://www.npr.org/2022/10/12/1127508725/prince-andy-warhol-supreme-court-copyright, accessed on 25th October, 2023.

13. N. Totenberg, Supreme Court copyright case looks at Andy Warhol series of Prince images on NPR, 2022, https://www.npr.org/2022/10/12/1127508725/prince-andy-warhol-supreme-court-copyright, accessed on 25th October, 2023.

14. https://commons.wikimedia.org/wiki/File:Triti,_Phanes,_625-600_BC,_Ionia_-_301224.jpg.

Networking – AKA Meeting People, Learning Things, and Making Connections

ANGELA ZIEBELL

Deakin University, Australia

Introduction

In this chapter, we will look at what networking is or isn't and see how you can develop your own approach to networking that suits your personality and experience level. We will help you get more comfortable networking, prepare for networking, and better understand the role networking can play in your professional development. There are a range of useful activities to do, either by yourself or with others, to help you prepare, including finding events that suit you. You will read about how networking can help you learn about other industries and careers, understand what others value in their company or in a colleague, and practice developing your own professional connections one by one.

There is a lot to think about in this chapter. It is very important to keep in mind that your brain can only take in so many new ideas at once and that this is the case for everyone (this is called working memory).[1] Just remember, this chapter is about ideas and tools. You should pick what you want to try, and what is most likely to work for you. If you are junior in your workplace but comfortable meeting new people, the ideas that you try from this chapter will likely be different from a second-year undergraduate student or a peer who is uncomfortable with large groups. Alternatively, as that same person becomes more experienced, they might start using new ideas that were originally more of a stretch for them. Just pick the ideas you feel you can work with for now and work on those. It is best not to try to do everything at once, as much as we often wish we could. Writing a plan of the top three things to work on over the next six months is a good way to stay focused.

"Just be open to everything. Yeah, just talk to people, go to events, try something new and it brings a lot of opportunities that you don't expect, and you don't know about."

– Ava, engineer and bio-nano scientist, former business founder and regional manager for a specialty engineering company.

Why Networking Probably Isn't What You Think It Is

Networking sometimes has a bad reputation. Some people have heard that it is all about who can do what favours for whom, it's for extroverts, or getting something you might not necessarily deserve just because you know someone. But this is generally not how networking works. If you are worried about these negative aspects, you might want to look at **Networking to Suit Your Personality** in this chapter to find out how everyone can find their own approach to networking and do it right for them. After you know how you are going to approach networking, then you might be in a better position to come back and start here again.

People (not just students and young graduates) often have the impression that networking is not real work, that it is all about getting favours, and that it is generally done by people who just want to get ahead. Networking can have a bad name and leave a bad taste in some people's mouths. But what if it wasn't that bad after all? Networking isn't really any more than getting to know people in your discipline area, sector, or type of work. These people can have very different training from you, but you might work on the same or related topics. You can meet people to share ideas, to learn about your industry and to hear what people are doing and why. You might share something that is useful to someone and *vice versa*. Just being up to date in itself is an advantage, but it is not an advantage that is unfair. You haven't taken from anyone to get an advantage; you are just getting informed. Will there be some people trying to be political? Sure. They are everywhere and they will be sure to attend an event seen as a networking opportunity. But they are unlikely to be genuine and that will hamper their ability to build professional relationships. Before we go further into this, let's think about some networking situations and think a bit more about why people network.

"I would say be open to learning about people and be genuinely interested in listening to them. Creating a genuine connection. Not just thinking this guy can get me something, right? It's very superficial if you want to do that."

– Joelene, Bachelor of Biomedical Science, now in marketing.

Why and When People Network

Any structured situation that brings people together (either online or in person) could be called a networking event. Most people don't really think about networking that way and usually think about big gatherings like conferences, staff events, guest talks or events that are actually marketed as networking opportunities. But most social encounters with a number of professionals in "work mode" can be considered a chance to network. Then there are also opportunities away from the workplace after work hours that can be very casual, like dinner with colleagues or a visitor. These are also networking opportunities. Both types of networking allow you to get to know people in your discipline or industry (or in neighbouring ones) and learn from each other. So, you could think of a training day, a meeting, a long walk down the street with a colleague, waiting for a meeting to start or standing in the coffee queue, as a chance for networking. Whether big or small, what do they offer? You guessed it, a chance to get to know people in your discipline or industry, or the related professionals that you work with or alongside, as well as a chance to learn what they know and what they are thinking.

> "Building networks while in university is crucial for building your preparedness."
>
> – Ava, engineer and bio-nano scientist, former business founder and regional manager for a specialty engineering company.

Some of you will be comfortable in some of these networking situations (like this casual lunchtime chat in Figure 4.1) and uncomfortable in others, so it's important to understand that they are all opportunities for networking. You can hate a crowd but be great talking with someone while you walk together, or great at being in the right place at the right time to have a carefully planned three-minute conversation with someone. In different environments (that could be because of the work culture or the culture of the country you are working in) these opportunities could be more or less available to you depending on your role, seniority or just who you are. As a vegetarian, I'm not going to go to the weekly Friday lunch steakhouse visit, and as someone who works from home, you won't hear the conversation in the hallways. But you know what you are comfortable with, and you know something about networking now, so you can start to experiment with what works for you. In the next section, I talk more about this, and you can explore more about what might work for you because of your personality and why.

Figure 4.1   Networking can be as simple as catching up with someone over coffee or lunch.

"I used to think networking was like, as soon as you're in a professional environment, you go up to talk to someone and that's that. I kind of learnt now, actually, networking is just literally what we do every day. And it's not even just going up to someone at a professional setting. I think to me, like my most successful moments where I've ended up networking have been almost something that happens indirectly as a consequence of just socializing. Like, I've spoken to people and I've ended up, you know, it ended up being a networking thing. Where like I've met someone at the dog park, while I'm walking my dog or just having a casual conversation. You start talking about different types of work and then you get people's numbers from that, or I've done this when I've been at a coffee shop. I've done this when talking to my friend's parents. Like a huge one that people don't even realise is like all of your friends have parents. Talk to them, find out what they're doing. Because a lot of, a lot of people get jobs through the [sic] networking.

And I think that's where people are like, well, "how do you network?" It's like, no, no, no. You literally just talk to people as you go on your day to day and you start to strike up a conversation like you would when you make a friend. You just start to incorporate more like what do you do for work?"

– Marina, psychology graduate and internship executive in a large corporation.

Networking to Suit Your Personality

Networking for Introverts

When people think about networking, they often think about needing to be talkative and extroverted to be able to succeed, but this isn't the case. What do most people mean when they use the word extroverted? Extroverted actually means that the person tends to perform their thinking out loud and needs to talk ideas through a lot as part of their thinking process. But in plain English, we often use the word to mean someone who is easily comfortable speaking up, who is comfortable in a setting with lots of different people, and it's often associated with people being outspoken and the opposite of shy.[3] The meaning of introverted is the opposite: someone who likes to think things through to themselves before sharing their ideas.[3] But again, that's not quite how we use it in everyday English. People in this position might be hesitant to answer, which can be interpreted as not having an opinion. They might have very strong opinions though, they just might not have finished all their thinking and be ready to share. They also might be unsure of how their thoughts will be received, or just have a habit of not talking much because they are uncomfortable talking in front of a group. We should also note that this is a continuum. People aren't just extroverts or just introverts, we all lie on a continuum between the two. Why have I taken all this time to explain this? Because understanding how people like to work, including yourself, is a critical workplace skill. It is also a part of your preparation for networking.

> *"Unfortunately, for introverts, like me, it is a challenge. I absolutely recognise that. But I think it's about networking to your strengths. So if you are introverted, you might start small, or it might be that you've gone to an event and there's one person that you talk to that then leads to having a coffee with them or a broader conversation or linking with them on LinkedIn, which can then lead to other opportunities. But yeah, it is important, but it's about it being genuine. So not sort of forcing yourself to be a different person because you think that this is what networking looks like. It's got to be genuine, because people can tell when it's not genuine or it's forced. And it is a really challenging position as a student or a recent grad to put yourself out there."*

– Alex, Bachelor of Arts/Science, master's in bioethics, lecturer in indigenous health.

It is important to remember that we are all comfortable in different spaces. Someone who is comfortable giving a technical talk and answering technical questions might not like 'small talk' (casual conversations about everyday matters such as the weather, sports, movies, what you are doing on your holidays, and so on). Or someone who can interact with people socially for a night might be really drained the next day. So in Activity 4.1, we will think a bit more about all the dimensions of where we are comfortable being social and why, to help us navigate networking.

Activity 4.1: Identifying the Best Networking Situations for You

Activity goal: To identify which types of networking situations you are comfortable in, which make you stressed and why, and therefore which are the best types of networking situations for you to start with.

Purpose and benefit: Making networking easier and more comfortable.

Activity steps:

1. Write down at least three to six different types of situations involving other people.
- On the left side, list situations that you avoid (the type where you would rather have a filling in your tooth than attend).
- On the right-hand side, list the types of situations you enjoy.
- In the middle, put those that don't interest you a lot but also don't worry you a lot.

These situations don't have to be entirely social, entirely work based or entirely learning based (for students). Just think of a mix of situations that involve information going back and forth and some aspect of communication and socialisation.

Some examples might be:
- going to the pub or a café with friends
- going to a restaurant with a wider circle of friends and their friends
- listening to a webinar, having the chance to ask a question, and then joining small group breakout rooms
- going to a product launch or industry seminar and hanging out with co-workers, eating the free food, and maybe meeting a few people someone knows from another company.

What other examples can you think of?
Which of these will cause you worry? Which ones are you comfortable with? Are there any that you can picture energising you? What might energise you?

2. Now that you have a few situations to think about, write down two to three things that come to mind that explain why you feel that way about each situation. (For this it will help for you to understand reflection covered

---

> ### Activity 4.1: Identifying the Best Networking Situations for You – (*Continued*)
>
>
>
> in Chapter 2: Transferable Skills and Reflection, and how important it is in developing yourself as a professional.) Work through this for the situation you find interesting (or at least not scary), the neutral situation, and the situation that you would not like to do. Make sure you think about what is behind your thoughts and reactions.
>
> For example:
> - If you are aware that it is the number of people there that you are concerned about, why is that a problem for you? You may be very clear that you do not like crowds and that's not particularly uncommon. But why do not you like crowds? How does that impact you?
> - If you are worried about too many unknowns in a particular situation, why do you think this worries you? Can you see a theme in the things that are worrying you?
>
> 3. Now do this with the situation(s) you feel good about. What is it that makes you gravitate to that situation, or at least what makes you think that it would be OK?
>
> 4. Lastly, go through each of the situations you have noted down and write two points as to what you are feeling when you think of yourself in that situation and why. Reflect on what is really causing each feeling. Again, there are no right or wrong answers.
>
> **Takeaway:** By now you might have realised a few things about how you feel about networking. Keep your notes and reflections handy as we are going to build on them later in the chapter.

Once you understand how you work, you will be able to pick situations that you are most suited for in order to practice networking. However, don't just sit entirely in your comfort zone. If you tend to talk too much when you are nervous, work on your listening skills. If you have a lot of trouble talking to anyone at all, start with talking more to people who are very "safe" first, such as friendly colleagues or a shop assistant who asks you how you are at the register. This is how you will build experience and gain skills and ability over time. The combination of understanding your comfort space and pushing your boundaries little by little is what will help.

For me, I like learning new things, so that helps me in wanting to go up to new and interesting people to ask them about their talk, their product, or an idea I heard them share earlier. But I still feel nervous. I still plan what I'm going to say beforehand, and I still get thrown off when I cannot find the right word, or I forget a name. And there are certainly some people I wouldn't feel comfortable just going up to, but much of that (for me) is whether they look like they will be receptive and/or are in a receptive group. So that's how I work now. But that's not how I was when I graduated. It took a lot of practice and I'm still very aware of some environments that make me uncomfortable.

> *"Networking is important, a challenge for introverts. If that is the case then network to your strengths, start small. It may begin with a conversation, a connection on LinkedIn, and then lead to more opportunities."*
>
> – Alex, graduate of a double degree in human rights and chemistry, previously a research assistant, now associate lecturer in indigenous health.

Getting More Comfortable With Networking Using Some Simple Preparation

Here I'm going to talk about a few things that might help you find your own style of networking. Next, I'll talk about getting prepared for networking. Those who find organisation to be their tool for handling challenges have already discovered that being really organised is very empowering. So, keep preparation in mind as an additional tool when networking.

It is important to remember that everyone is different. Having said that, I find the single biggest concern with networking is people not knowing what to say or worrying about what to say (OK, technically two points). So, let's think about this for a moment. First, even though individuals sit on a spectrum of what they prefer, people generally love it when you listen. It is human nature to like to talk about the things you are interested in, and to have someone listen intently, or at least be clearly interested in what you are saying. The benefit here is that you have time to listen and digest and learn a lot about this person, what they are doing, what their company is doing and why, what the trends are in their industry, *etc.* It is an information goldmine for those of you trying to enter the workforce or navigate those first years out in the workforce.

Second, you can use the same questions on most people. Jan does not know that you just used the same four questions on Ali and Jez! So, you do not actually have to be saying a lot or asking a lot of original questions. That does not mean there is an excuse for being lazy and just relying on making things up as you go along. For example, don't ask the speaker how they liked the talk!

For some of you, working things out as you go along will make you feel stressed. Others are stressed at the idea of how to prepare. Let me give all of you some tools to start with. So let's sit down and plan like the person in Figure 4.2 with a brainstorming activity, Activity 4.2.

Figure 4.2   Plan out how you can get more comfortable with networking, including what types of networking work for you.

Activity 4.2: Brainstorming Questions for Networking (Repeat When Your Audience Changes or You Need Better Questions)

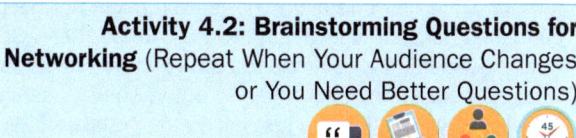

Activity goal: Write seven questions that you can ask most people, most of the time.

Purpose and benefit: You are writing questions that are suitable for asking almost everyone, almost all the time. You will still have to listen and pick the appropriate question from your list, of course. But an average respectful amount of attention is all that is required. By having about seven questions worked out, tested and loosely remembered (it does not need to be word for word), you will be empowered to add to the conversation, or restart an awkwardly stopped conversation. And importantly, this will allow you to relax a little.

Activity steps:

Guiding principles for writing and using your questions:

1. Make sure your questions have a professional side to them, but you can also feel free to put something in that is not particularly about work but still has a professional tone. For example, if there is a holiday or major event coming up you can ask someone what they are doing to celebrate the event. This question is not anything to do with their work, but still involves a professional tone and topic.

Continued

Activity 4.2: Brainstorming Questions for Networking (Repeat When Your Audience Changes or You Need Better Questions) – (*Continued*)

2. When writing your questions, make sure they are open questions. An open question is one that is impossible to answer with just one or two words. Yes/no questions are bad for making conversation, so get people saying more by using an open question. For example, I could ask someone how long they have been at their workplace. While technically not a yes/no answer, the answer could be really short. If I instead ask them "how long have you been at the company and what changes have you seen at the company since you joined?", then I'll almost certainly get a longer answer.

There are some obvious follow up questions to this example too. Do you think the changes are for the good or the bad? What do you see the impact being on the company? Do you think any of these are part of a wider trend? If they gave a really detailed answer, you might not need these follow up questions because they might have answered them. So be sure not to just remember a list of questions to ask. You need to be paying attention to what they are saying in order to select the next question. If you are thinking, "what if they say they only just joined the company?" Well, all the above still applies but you ask about where they were last. If they happen to have just graduated, well, ask them what course they did and why, *etc.*

Testing your questions:

1. Now that you have your seven questions, think about a few different circumstances and check if you would be able to ask almost all of the people there most of the questions, most of the time. Think also about whether the questions cover a variety of topics. If your questions are all on similar topics and the person you are talking to seems uninterested in talking about those topics, then you will be immediately out of questions. So, devise questions that are spread over a variety of interests and topics, and check if they could be applied to a range of disciplines and businesses/organisations.

2. The next step is to find a friend or family member that you can practice your questions with and who can give you some feedback. For this, it is ideal if they are also interested in or familiar with networking, whether this is because they are just starting out, or are more experienced. When you are talking, see whether their answers to your questions naturally prompt more conversation and questions. If so, this is a good sign.

Context: If they are a professional, be sure to consider the environment that they network in and ask them what they mean by networking and why. Networking can look very different in different specialties or disciplines.

Activity 4.2: Brainstorming Questions for Networking (Repeat When Your Audience Changes or You Need Better Questions) – (**Continued**)

For example, teachers would usually do most of their training and interact mostly with other teachers. However, medical professionals work daily with a very mixed group of professionals (e.g., nurses, physiotherapists, specialist doctors, social workers, family doctors, and testing laboratory personnel). If you are going to get feedback, it is crucial to understand where the other person is coming from in terms of background and experience, so you can fully take advantage of this process and their thoughts.

When thinking about how it went when talking to someone else about your questions, or trialling them on someone, think about how they answered. How rich were the answers? Did the questions give you enough information to create further discussion around? Did the questions start and sustain a conversation? Did they give you something to ask more questions about?

Takeaway: You are not responsible for keeping the whole conversation going. But with a little bit of preparation (and maybe a colleague, friend, or classmate for company), you can be ready to add to the conversation and answer a few questions when they come your way. While you won't progress immediately to networking genius, don't underestimate how much good preparation can help.

Hint: When "remembering" your questions, you want to know them well, so you don't have to worry about forgetting them or stumbling over your words. However, be careful not to share them automatically, a bit like a robot. You tend not to be thinking about the interactions and what people are saying when you have rote learnt questions. It can sound really scripted and risks things like asking a question that has already been answered. Or if you lose your place during the question, you might not remember the next word. So, focus on remembering the general idea of each question, not the exact wording.

Talking About Non-work Topics

If you are new to networking, you might not know that non-work topics often take up just as much of the conversation as work topics. This can vary quite a lot, so my best advice is to follow the average approach that others are taking. You will be fine as long as you use polite language, and don't stray too much from the types of conversation topics everyone else around you is talking about. It is actually common that people new to networking find this uncomfortable, as they are working on the assumption that networking is all about work and that it is unprofessional to talk about non-work topics.

Some people are much less comfortable talking about personal topics, or are just very work focused, so again, take your cues from what others are asking you. Can you identify who looks uncomfortable in Figure 4.3? If someone mentioned a hobby, pet, or family member explicitly, this is a cue that they are happy to talk about these things and so you can enquire further about that topic. For example, how many pets do you have? What type? Or, how old are your children (assuming they mentioned children)? Relationships are often formed around these non-work topics. You might find someone who shares a similar hobby or supports the same sports team as you, has a passion for baking like you do, speaks a language you speak, or grew up in a nearby area to where you grew up. Any point like this could end up being central to a relationship that turns into a friendship. I once worked with another vegetarian my age who also had a passion for cooking and growing fruit and vegetables in her garden. I left that workplace 15 years ago, but I still go back every year to have lunch with her and we talk about our new recipes.

Figure 4.3   It is usually appropriate to talk about non-work topics. If you are unsure, take the lead from the people who you are speaking to.

Why Network? The Benefits of Networking for Experienced People and Students/Graduates

> *"Network with other graduates, whether they've gone down a graduate pathway or whether they haven't gone down the traditional graduate pathway, someone that started their own business, people that are working in startups and in that startup space. Yeah, talking to a range of people, I wish I did that more."*
>
> – Nigel, Bachelor of Science in physiology, former business analyst and COVID-19 team leader, currently a marketing specialist.

You might find it a bit hard to believe that someone junior like yourself could have anything much to add to a conversation, whether you are still a student, or in the first few years of your career. However, there are a number of very important reasons why someone who is more senior might want to talk to you. And no, the reason is not pity. The following are all very real reasons why more experienced people like to talk to students and junior employees. Read further for details on each.

- You are the future workforce,
- understanding who is learning what,
- to build connections generally,
- learning from others,
- giving back to others or the industry in general,
- getting to know people.

> *"I look at my networks and my networking skills have taken me very far. It's my network which has always helped me achieve everything."*
>
> – Shezmin, previously a research assistant in microbiology and chemistry, now developing new applications for new specialty equipment.

You Are the Future Workforce

First, it is very important to remember that students and young workers are future potential employees. It is very sensible to get to know the people that you might end up working with one day. This is not necessarily either an altruistic move or a political move, it is just common sense. If I am a senior person, I know I might need someone for a job in the future and it is helpful to know something about the people I might like to hire.

In fact, I might find someone who I think is great and an excellent match for my team, and that will make the hiring process so much easier next time. I can also give a lovely person whom I respect a job, or maybe suggest them to a colleague or friend. Hiring people is hugely time consuming. Sometimes there are thousands of applicants for one or two positions. If I'm in the above position, I save everyone time if I can go straight to some-one who is a good candidate, or even a handful of people who are good candidates. Lots of jobs are informed through these casual interactions. As a student/graduate, you aren't cheating by making sure that a potential future employer knows that you are competent and, for example, have val-ues and a skill set that aligns with what the company/organisation needs. However, if you don't go out and meet a few people (*i.e.*, network!) and practice talking to these professionals, then they can't think of you in the future when a job arises or know how amazing you are.

Understanding Who Is Learning What

Second, people in your discipline or your target industry are often interested in the types of things you are learning through your studies (or learnt, if you have finished). These people are interested in keeping up to date with what is happening in the lives of young professionals. Would you hire the person from the university that trains in the exact set of skills that your company needs? Or would you hire from the university you know does not quite align with your industry? For a candidate that is otherwise equal, an employer will hire the one with the degree that is most aligned. But the exact content of the degree is not that obvious, even from the course description, so talking to the students lets someone ask really specific questions to get more detail. At the same time, they might be also determining how much you understood about your course too, which would feed back into the first point: getting to know those coming into their discipline/industry. It is not an interrogation, but if you cannot explain what was, or is, in your course, then there would be some more questions that I would want to ask, to see if you had the right skills. At this point, understanding what you have learnt, why you learnt it, and how to talk about it is key (refer back to Chapter 2: Transferable Skills and Reflection if you need further help to understand how to explain your skills to others as well as how to fully understand the skills you yourself have developed). This is often underappreciated by young people.

Building Connections Generally

My third point relates to building connections generally. There are a lot of reasons for networking. People often network to learn incidental things and build professional connections that might lead to something in the future, such as to collaborating on a future project or idea. This is very vague, so let me give you an example. The co-editor and I met when we were both teaching in the same class part-time while we were also doing other things. We kept in touch because we got along and over the years we talked about internships and possible university practical exercise ideas. Then a few years ago, I invited Rebecca to work on a session of my career skills class to give students a chance to talk to someone who had run their own business. (She ran a biofuel start up at the time.) I wouldn't have known who to bring in if I hadn't kept in touch (networking). We had coffee a few times because we like to stay in touch (networking). Now we are editing a book together and writing some of the chapters. We first met over 9 years ago. I had no clue when Rebecca asked, "Do you want a kilo of chocolate from the machinery I bought for my new factory?" that we would be working together 9 years later. But that's how these incidental networking relationships work. Nothing happens if you don't keep connections going, but also you don't usually keep them going with a specific professional goal. You keep them going because eventually something great often happens, professional or personal or maybe both.

"It's so important... life is literally about who you know."

– Joelene, medical biosciences graduate working in marketing.

Figure 4.4 represents a group of friends from university. Who will you stay in touch with and how? Or who will you get back in touch with?

Figure 4.4   You can never have too many good quality connections. These can just be your peers that you stay in touch with over time as you all progress through your respective careers.

Learning From Others

This leads firmly to my last point. Through networking, you will learn from each other. This is something that we all already do as students. You hear someone explaining a difficult concept that they have just worked out, and that helps you understand. Or your classmate hears a piece of news that is useful for you to know, and you file it away or you share some useful information with another student. For example, maybe another student shares that the exam in the class you are completing next was hard but fair and needed lots of preparation. In this case, you might use that information to spend more time preparing. If you know that person well, you might have additional insights. For example, if you know the person is terrible at preparing, then you might conclude you will be fine. If we actively seek to listen to those around us and to ask thoughtful questions, we can learn a lot. And remember, it might be information that you don't even know you will use yet. It doesn't have to be immediately apparent how it will be useful, to become useful later. All the knowledge you gain as a young professional is giving you a better background knowledge of your industry or sector.

As you build relationships, you will also form a group of people who you can ask useful questions. As a teacher at university, I might like to ask a high school teacher a question about their curriculum. I can't just do this if I don't know a relevant teacher whose thoughts I trust. Networking means that you develop, over time, a set of connections that you can learn from in the future.

Giving Back to Others or the Industry in General

Through networking, you will also end up helping each other. These are not the type of favours where you help people only so that others miss out (although that is what some people picture when they think about networking). The type of helping I'm talking about here are things such as:

- you mention a role is available to someone you know who might be interested
- you help to connect people you think might have something to talk about
- you recommend training or a resource like a book or paper
- you suggest ways someone might deal with a problem they have.

These sorts of exchanges don't come along as a formal deal where you know you will get something specific back. They don't need to. These exchanges are just about helping where and when you can and being thoughtful about others. This is not to say that you can't then ask for help when you need it. Again, this is not about "getting things" from people;

most people are happy to help. Think about yourself. On average, if someone asks for a suggestion or you know a resource that might help someone, do you help? Of course, there are occasional exceptions (*e.g.,* maybe you don't know what information you are allowed to share), but in most cases, most people do help.

There are a few key things to keep in mind though when reaching out for help. First, how established is the relationship and how is that relationship progressing? If it is not a significant relationship, then you cannot just email someone and ask for a personal recommendation for a particular role at their company. That would not be appropriate because you don't know each other very well. Alternatively, if the same person has expressed interest in you being part of their company (*e.g.,* "I think our company could use more of your skills" or "I think you would fit in really well at our company"), then you might talk to them. To tread softly (if that is your style or if that suits the relationship), you might make a chance for a conversation (phone, Zoom or email if you cannot meet) and talk to them about how you are thinking of applying to a role at their company and ask them if they have any advice about your application. Then you can largely let them take the lead and see what they offer. Take into consideration the person's personality, or what you know of it at least. People can respond better to a more direct approach or a less direct approach, depending on who they are. This is where the power of networking comes in. The better you know people, the better you know how to talk to them and what they are interested in.

"If you don't have a network around you, you won't ever find out about certain opportunities. You won't have people to really vouch for you. So I don't like networking for the sake of networking, but in terms of you bringing value to other people and then by extension, other people being able to bring value to you, it's super useful... Always be thinking about the value that you can bring to other people. If you see an opportunity that you think would suit somebody, recommend them for it. If you see a club at Uni that needs the skills that you can provide, go volunteer for that thing and just be a nice person. That's pretty much it."

– Dylan, Bachelor of Arts in psychology, working in finance and doing a master's in data science.

Getting to Know People

Last, but very importantly, networking gives others the opportunity to get to know you. If this immediately seems scary, I suggest you read The

Spotlight Effect[2] in the following section, to help put your fears into perspective. This section will also help you with concerns related to making a bad impression by not knowing what to say or not being able to answer questions on the spot.

Whether you are comfortable with the idea that it is beneficial to let other professionals get to know you or not, it is beneficial. Think about this. Can I offer you advice if I've never met you? Can I recommend training if I don't know what your needs are? The answer is, of course, no. So, let us think about the benefits of letting people in your professional community get to know you. First, neither of the above points is possible if no one in your professional community knows who you are and what you are interested in. Once you start to get to know people, they know who you are, and you can start to develop relationships with them. Some will just be familiar with your face and honestly, that is still OK. If you walk into a meeting one day and someone there recognises you, this will enable you to start talking to them and get introduced to a number of other people at the meeting. That is a real positive. Familiarity is a great thing and people feel safer choosing someone/something they know. So, people simply being familiar with us can be a plus.

> *"People hire people they like and they know."*
>
> – Thomas, business founder, environmental scientist.

By getting to know a range of people and by allowing them to get to know you, you might walk into a new workplace one day and see someone that you have talked to a number of times. Or you might know who to email when you are looking to learn more about a particular industry. You might think that this would be a rare coincidence, but if you are networking in the same area/industry/profession that you want to work in, and taking time to talk to the people at the companies you are most likely to apply for, this is not such a big coincidence. If you are still worried that having met someone before just gives you more time to make a bad impression, please remember The Spotlight Effect discussed in the next section.

Are you able to invite someone you respect to sit down and have a one-on-one chat like we see in Figure 4.5? If not, think about why. What would help you get there? Or maybe you would feel better about practicing online first.

Figure 4.5   People can't help you or give you advice if they don't get to know you.

Ultimately, networking connects you with a range of people who are in your discipline or field. You might have a lot in common with some of these people and you will enjoy the company of some of them. However the relationship starts out, over time you will probably end up with some friends that share an understanding of your work. And that is a good thing to have!

The Spotlight Effect

On average, people think about you and what you are doing about 6 times less than what we think (Mendoza-Denton, 2012).[4] Let that sink in. For every time you think you did something silly, and that you think someone noticed and remembered, there is only a 1 in 6 chance that they actually did. And that's on average. If you are particularly concerned and focused on what you did, then the number of people that noticed what happened might be one tenth or one twentieth of what you fear it is. That's only a 10% chance someone was paying attention to whatever you did or said that you are mortified about. This is called the spotlight effect. When we speak or act, we feel like the spotlight is on us. This is a totally natural thing to feel, but we need to be conscious about the impacts of the spotlight effect. When you are in a networking environment and feeling self-conscious, keep this thought close at hand. People do not pay as much attention to you as you think they do.

It is also important to remember that we tend to overestimate how much other people care about slight mistakes that we make. If you "um" and "ah" a couple of times (or even more) when you are trying to remember something, that's normal. Mis-remembering that Anthony goes by Tony is not something people really care about that much. Of course, it is worth putting effort into trying to remember these things. Being thoughtful and respectful of who you interact with is a very important skill. But people's mistakes or slight missteps are not really that memorable. Would you remember the person who said an "um" just because they said an "um"? Instead, when we are nervous or self-conscious, all these things usually seem much bigger than they actually are.

So, what can you do about being self-conscious? It is helpful to accept that you might feel self-conscious in certain environments. It's important to remember that you are certainly not the only one that feels this way. There are many people, even those experienced at networking, who feel self-conscious. They might be hiding it well so you cannot tell, but certainly some experienced professionals still get nervous when meeting new people. Alternatively, some experienced professionals are relaxed now, but most were not when they were a student or new graduate. It may have taken significant experience and/or effort for them to become comfortable with meeting new people and keeping conversations going with people they have just met.

So, in summary:

- people don't really notice everything little thing you do, so relax,
- people do not really care if you make small mistakes,
- other people are self-conscious too,
- and practice will help.

Strategies for Preparing to Network

So, by now you should have a list of great networking questions all sorted out. Maybe even a few spares just in case it is a long day. And you have also practiced with some friends and family. You even know all four main points from the previous section. Brilliant. What next?

While it's not good to worry about networking events for days on end, some simple pre-planning when thinking about going to a networking opportunity can go a long way.

In simple terms, you need to think about where, who, and why. First let us look at the who because this has two parts. Who are you going to go and see (this could be a general "who" like a company, or it could be an individual) and who are you going to go with?

Who to Go With

While you don't have to go with anyone, I think most people new to networking have more fun and feel a lot more comfortable when they go with someone they know. Importantly, when events are at night, this can also minimise safety concerns about travelling alone. For those of you that do not have this concern, great, but that does not mean others do not have this concern. Therefore, asking who would like to go together can help you help a classmate or colleague feel safer attending. And you know what? That would be a great step in building a relationship with that person – another example of helping as a way of networking (see above – Giving Back to Others or the Industry in General). Refer to Activity 4.3 to work out who to invite and how.

Activity 4.3: Identifying Who to Invite to Come to a Networking Event With You

Activity goal: Find someone to go to your next networking event with.

Purpose and benefit: Networking can seem overwhelming and even when you are more experienced at networking, a specific event might worry you more. Find someone you know who wants to come too and support each other.

Activity steps:

Having trouble working out who to ask about attending a networking event with you?

Try the following. This can apply when you are asking about a specific event or just generally asking if people might be interested in coming to the next event, if the timing works for them.

1. Start with a list of people who just pop into your head for whom the event might be relevant (classmates, colleagues, friends, people in the same clubs, *etc.*).

2. When talking to the people from point 1, ask them if they know anyone else who might like to join you.

3. If you are comfortable, post on social media to ask who else is going or even just a post about looking forward to meeting people when you are at the event.

4. Once you have found someone, be the one to organise where to meet and when, *etc.* It is amazing how people can forget, or if they don't

Continued

Activity 4.3: Identifying Who to Invite to Come to a Networking Event With You – (*Continued*)

see an email close to the date think that it isn't on anymore, *etc.* Put it in everyone's calendar, watch for acceptance notifications and remind people if they haven't accepted. This is great professional practice, and many junior people haven't learnt these skills. Your classmates and colleagues might need a bit of help understanding diary software and the differences between all the different platforms for email.

5. Check who is still going close to the event. This is a polite reminder and also a time that you can express your enthusiasm, something that will help people feel like this is something worthwhile to attend.

Takeaway: Going to an event with someone can make everything feel more manageable. From organising yourself to planning who to talk to, from practicing asking people questions to what you might say if you are asked a question, preparing with someone else is going to help.

Figure 4.6   Organising to go to a networking event with a friend or classmate can make going to events a lot more manageable. You will likely feel more confident and get more out of the event.

As for who to go with, just ask around in advance and see if anyone that you know at work, from class, from a university club, from work, *etc.*, might like to travel together or meet up at the event. The two students in Figure 4.6 genuinely represent how working with a friend can help with networking. It's work but it can be fun too. Revisiting the point from above that lots of people will be nervous about attending, you may get more interest than you expect, as once they know they have someone to go with, more people are likely to decide to come. Just knowing that you are meeting up with someone for a chat at a particular time during an event might help to put your mind at peace a little. So, if you can, find someone to meet up with or spend some or most of the event with, and you will have company and not feel so self-conscious. This is a really great approach, and even seasoned networkers will do this.

However, a word of warning is not to spend most of the night just talking to the person or people you came with. Find a couple of people to introduce yourselves to. There will be at least two of you, so you can talk about whom to approach and if one of you stumbles for words, the other can pick up the conversation and keep it going. Remember the questions you built earlier and use those. It's also important to realise that no one at a networking event wants to be standing alone with no one to talk to. So, it can be a great strategy to go together to talk to someone who is currently standing alone. They will likely be pleased to talk to you.

Who Is Going to Be at the Event?

The other "who" that it's important to know a bit about is who is going to attend the event. Remember, we are talking here about all forms of networking so they could be formal, planned environments or they could be a casual lunch with a group of visitors.

What you know about the "who" can be very different from one context to the next. If you are lucky, there will be a full list of guests with company affiliations and biographies on a website or a flyer. Other times, it might be a bit harder to find out the details of who is going. Sometimes you won't really know and that's OK too. You might just be able to research the companies. This is still a great way to prepare so you know better who to talk to and how to approach them based on the company they are from. However, no matter the situation, if you start with some preparatory information gathering, you will do better than if you do not gather information beforehand. For example, you will know who you are most interested in talking to, and about what. You can prepare some specific questions for them (*e.g.*, about a role they have had, or their industry/ company), or plan points to help keep a conversation going if you are able to talk to them (*e.g.*, what was it like to work at ___?, how big is ____ company, and what skills do they usually need?). I find questions are the easiest form of communication when you are trying to talk to someone you have never met. This is because you are tapping into the idea discussed earlier that asking questions (that you can pre-prepare) is easier than leading the discussion yourself.

Getting the Most Out of Your Time at an Event

All this preparation will help you gain the maximum benefit for the short amount of time you are at an event. This will also reduce the amount of time you invest in talking to someone who was never going to be of interest to you. For example, maybe there are two representatives from a company that you are interested in knowing more about. But one representative is a senior scientist, and the other is an HR representative attending to promote an internship program. If you are a student and mostly want to hear about the internship, then you will want to talk to the HR representative. If you are a graduate wanting to hear about what technical challenges the company or industry faces, you are going to need to talk to the scientist.

As you start to develop some connections, you can talk to your relevant connections to prepare for the event. Does your boss or a student you know that graduated in the year before you know anything about the events or companies? Who has been to this event before that you might talk to? Or even a similar event. Do you know someone who knows someone, that might know? You can ask people to find out these things if you are polite and keep requests to a minimum.

For example, my cousin is a young engineer interested in data and artificial intelligence. If I ask him if he knows if anyone from a competitor's company is likely to be hiring at the moment, he might know. And if he does not, he will be happy to ask. And... did you spot it? This is networking! Now I am not particularly close to this cousin, but he is lovely, and I hope he knows that I would help him in this sort of way, too. So, it definitely passes the test of whether or not it is OK to ask for this information. But, if I did it every second week, he would not be as helpful. That would be terrible networking. And let us not be that cousin that always asks for favours.

Researching Individuals Before an Event

To prepare for an event, do what these seasoned professionals are doing in Figure 4.7 and sit down to make a plan to help you approach the event. It can be even more helpful if you can sit with others for feedback and ideas, too. There can be special guests and speakers at some networking events, which you can usually easily find out about beforehand. These special guests are easily identified from correspondence

Figure 4.7 Before an event, your secret weapon is research. It will allow you to plan interactions and be more confident making conversation, both with specific people you might want to talk to but also in general.

about the event, or the event website usually comes with a photograph, or the guests are well-known enough to have an obvious company picture on the internet that will be easy to find with a quick internet search. They will also almost always have a description of themselves, their career and sometimes their company on the event page. Or if not, again usually something will be available with a quick internet search. With more and more video available on the internet, there is also now the chance to watch the guests talk beforehand. There may be a video from another event, or a more formal news piece, or a YouTube video, or maybe a company promotional video. This is a fast way of learning about people. What do they care about? What did they choose to speak about? What are some important points for them within those issues? Who do they seem to be as a person? Using any of these different pieces of material for a relatively quick preparation, it can be much easier to be ready to talk to these people on the night. You will never be ready, you say? Well, now you can be closer to ready.

Alternatively, if you happen to watch the person speak beforehand and they are not of interest, then you know not to worry about working up the courage to talk to them at the event. Instead, you can spend your time preparing to talk to one of the other guests. I once met someone whose work I admired greatly and they are greatly acclaimed, but I did not have an enjoyable conversation. And that person did not seem to be that interested in talking to me. Now that story is a great anecdote because this was a Nobel Prize-winning scientist, and I realised that I did not particularly want to talk to them. Until then I had held such people on a pedestal. In a way it was a shame, but on the other hand, this can happen, and it is important not to take it personally. This is an important learning step in networking. There are amazing and interesting people to meet everywhere. Sometimes the high-status people are not who you need to be talking to, and they are certainly only a small proportion of the amazing people you can meet.

Researching Companies Before an Event

As well as individuals, there are a lot of ways to find out about companies once you know what companies might be at an event. The obvious place to go is the company's website, but there are also the company's social media accounts (you can follow on LinkedIn) and any news or general internet pieces that might be online. As with individuals, the company might also have videos online, either about the company themselves and what they do, or possibly about an important issue in their industry. The company might work to influence policy changes, for example, or they might support a charity with income, goods, or by allowing their staff to volunteer. Occasionally you might find negative press about a company too. For example, they have been part of a lawsuit, they breached an environmental regulation, or they are treating their staff badly. Both positive and negative pieces on the internet need to be looked at with a critical eye. For example, how many sources agree, or is the case ongoing, or was there a finding? As STEM students/graduates, you are well equipped to navigate such issues. The point here is that there is a lot of information out there about many companies or about any given industry in general. There is a lot to learn and looking at companies when they appear at events is a good way to learn about them. Those companies may be potential employers for you, as they are active in your region. Work through how to prepare for an event in Activity 4.4 below.

Activity 4.4: Preparing for a Networking Event

Activity goal: Find a networking event and prepare for it.

Purpose and benefit: Learn how to prepare for a networking event, including learning about the companies that will be there. You will be much more prepared once you arrive at the networking event, and you will be able to make better contacts and more easily have productive discussions.

Activity 4.4: Preparing for a Networking Event – (*Continued*)

Activity steps:

1. Find an event that might be good for networking.

2. Research the event to find three companies that will be there (you can call or email the organisers if the guests are not apparent, but usually they will be).

3. Then research these three companies to try to find answers to the points below. See what you can find out.
   a) Research what the company does, who owns the company and when the company was founded.
   b) Find four points about each company that you did not know before.
   c) Make sure one of those points is interesting enough that someone outside of the industry would find it interesting. This will arm you with a piece of information that can help start or extend a conversation, whether at this specific event or at another event. Alternatively, there will be people at the event that are less connected to the discipline, for example HR representatives. A generally interesting fact will be a conversation starter for people at the event that don't share the same technical background.
   d) Check to see if they have any jobs open at the moment, either that fit your interests or not.
   e) Find someone at a similar point in their study or career (can be the same discipline or different) that will join you to work on this research together and compare notes when you complete step three. (If you would prefer to do this part alone that is fine, it can just be very helpful to get external ideas and feedback when preparing.)
   f) If you cannot find anyone to do the exercise with, find a friend or family member to talk about what you learnt and see what they think. Ask if they would like to know anything else about the company before they talked to someone from the company.
   g) Is there something more you would like to know about this company after your initial research? If so, note down these points/questions. There may be an opportunity for you to ask these questions from their representatives at the event. This will help you learn more, as well as demonstrate your interest.

It's very important to learn something about different parts of the company, not just your discipline. This will help you better understand how the whole organisation works and why. Much more about this in Chapter 3: Commercial Awareness.

Takeaway: Once at the event, the above will give you a level of understanding of the company that is enough to have a conversation with someone from that company. Combined with your seven questions for networking, you will be all set to join or start a conversation.

It is important to remember that the host and special guests are generally there to help the event succeed. You have to be respectful of their time at the event, however, they are perfectly placed to answer some questions and potentially introduce you to someone else at the event. In accepting the role of being the host or the special guest, they have generally also accepted that they will socialise with people at the event and help the event run well. If the event is anything related to careers or training, then they will also be expecting a lot of younger or more junior people, and they will be very happy to talk to you. In most cases, it is literally why they are there.

How to Spot Who to Talk to and Who Not to Talk to

There are always guests that you might be better off not talking to and you won't always know that beforehand. And, while you might get a little bored talking to the person who only wants to talk about their work in an area you are not interested in, this is not a negative interaction. It can be quite useful to spend a bit of time finding out what people in other areas do. All learning can inform you later. However, some interactions can leave you feeling discouraged and deflated, and these are the ones I want to mention here. People come in all personalities, and we all have good and bad days, so it is important to practice observing someone's body language when they are talking to others and work out if they are a good fit for the type of interaction you feel comfortable with. Does the person take over the conversation? Is this a positive or a negative for you at this time? Is the person combative or loud? Is this a good fit for you? I think we all know we do not want to spend time with someone that is obviously rude. But it is not just as simple as avoiding the one rude person. I also think that your success in feeling comfortable at an event is strongly influenced by each individual interaction. If it is your first event, seek out the friendly people that you see being obviously friendly to quieter people, have a chat, and ask them to introduce you to someone. Feel free to also ask to be introduced to a certain person or a certain type of person, too. You could, for example, ask if the person knows anyone in your specific sub-field or in a certain role. It will actually make it easier for the person you are asking to think about someone, and certainly make it more likely that you get introduced to someone who you will learn a lot from talking to.

What Happens When You Start Talking to Someone and Then Realise It Is Not a Good Fit for You?

This can go either way. The person might be super quiet and the conversation not going anywhere, or the person could be a bit much for you and you do not feel comfortable. Most of us will stress a lot about how to end a conversation. Will we cause offence? Will it be awkward? Will they think you are weird? This is the spotlight effect coming out again. When you leave a conversation, you do really have to do it spectacularly badly for it to leave a lasting negative impression. People do not really notice it anywhere near as much as you think (times six, remember). So, you could say you need a drink or some more water, or you have to make a phone call and politely step out of the conversation. If the person is very senior, then there might be some extra difficulty in your mind whether you should leave the conversation. You should, of course, always respect the local or company culture with regards to how you end a conversation with someone more senior to you. You also need to take their personality into account and whether they have influence over your job. Someone senior that you might not see again and seems nice is very different to a notoriously grumpy boss.

In the above case, can you make it their idea to start another conversation? For instance, if they were talking about how you should meet someone and you see the "someone", then you can remind them of their idea and ask if now would be a good time to make the introduction. Or maybe they mention a display, or a brochure *etc.*, and you can ask if now is a good time to look at it. If it is their idea, then you are not being impolite. Or maybe you can offer to get them some food or drink. They may well be busy with someone else when you get back. Remember, the aim is to give you tools so that you can feel comfortable navigating awkward situations. These situations might not even happen, but we tend to worry about them. So knowing some potential solutions is a great way to feel better about the thought that this awkward situation might arise. This might happen online too. Don't be scared to be the young graduate from Figure 4.8 politely having to say, "well this has been nice, but I have to go now". Activity 4.5 can help you have a plan for removing yourself from an uncomfortable conversation. But remember, these aren't that common.

Figure 4.8 Either online or in person, conversations can get unproductive. You need to develop ways to exit these conversations constructively.

Activity 4.5: Removing Yourself From an Uncomfortable Conversation

Activity goal: Have a list of polite ways to excuse yourself from conversation to empower you to navigate conversations more skilfully.

Purpose and benefit: The more tools we develop to handle situations that might worry us, the more in control or comfortable we can feel. This will help you feel comfortable where you otherwise might not, and allow you to relax and participate in a lower-stress manner. You might also upskill enough that you find yourself being able to go to events that otherwise would be outside of your comfort zone.

Activity steps:

1. Take a moment now to picture someone specific that you know of that you might have this problem with. You do not have to know them well, just a bit about them, even if that might just be *via* webinars, word of mouth, or television. Maybe you have heard they are boring, or grumpy or very loud and you worry that you will be stuck talking with them and not know what to do.

2. Think about the type of ideas I just went through above. What would be the three things you would do to very politely remove yourself from a conversation with this person that you felt had gone on for way too long? It certainly does not have to be from my list of suggestions, these are just starting points.

3. Take these three points and talk to friends and family to see if you can fine tune or add ideas. Write down your final plans and keep them in mind if you start to worry this might happen.

Takeaway: With a little planning you can have a few reliable approaches to most conversations that you want to end for whatever that reason might be.

Tips and Tricks for Those Uncomfortable in Many Networking Situations

Here we are going to talk about some general tips and tricks, but first let us think about when we talked a little about introverts and extroverts. This included both the official definitions and how people often use these words colloquially. Whether or not you are on average an extrovert (whichever meaning you use) does not provide a lot of "protection" if you do not feel comfortable in a networking situation (either formal or informal). So, for a minute, let us all think from the perspective of someone who is uncomfortable in many networking environments. And remember, this could be anyone, not just an introvert. This person might become a real networking champion over time with thought and practice (again, this could be an introvert or an extrovert). This person is not bad at networking, and it is actually really important to remember this. Networking is definitely something you can improve; it is not a fixed state. Think about how much your writing has improved since you were five years old. Does anyone believe that a five-year-old's writing will not improve? Of course not. And it is the same with networking. People will generally always get better if they do something, even if they are not working particularly hard at it. If you are working hard at it, you will get much better and work out your own style, what sort of events are best for you and why, as well as how to prepare.

You will work out the best way to navigate the occasion given what you know and what you have practiced above. In the future you might think, "I should have done that differently", and you learn another way to ask a question or prepare for meeting someone. This is great reflecting (see Chapter 2 on reflection). But we need to keep reflection and worry in balance. Let us assume that, as you go through this process of learning how to network and what works for you, you are using all of what you know at the time, and you will get better over time.

For Someone Who Is Worried About Saying the Wrong Thing

The best idea is often to keep the other person talking. This is where you take out those seven questions you designed to bring to the networking situation. If the other person is generally doing all the talking, then you are free to listen and learn, or even plan your next question (just make sure you do not lose track of the conversation). You can ask them if you can note down certain things that they mention that you think are particularly worthwhile. But when taking notes, be sure to make it clear you are still listening so as not to seem rude.

Why does this work? Well, honestly, everyone likes to be respectfully and attentively listened to. And the vast majority of people like to talk about themselves and their work or topics that interest them. Or maybe even their last holiday. Having a polite and friendly audience can be a very pleasant experience. Be sure to keep track of what you have asked to whom and what they have answered. You need to keep thinking while you listen. Having a list of questions prepared beforehand is not the point at which you stop thinking. Do not, for example, ask the guest speaker, "What brought you here?". They clearly attended to be the speaker. I suspect most of us have done something like this once. Remember, people really do not care as much are you think they do. But it is, of course, best to try to pay attention and not make such a mistake. Instead, you just need to fine tune your question and ask something like "Have you been to this event before? How did you come to be there the first time?" or "Who invited you and how do you know them?".

You will not be free from the obligation of answering their questions, of course. You need to expect that every seasoned networker will use the easiest question of all to ask. That is, they will just ask the question that was last asked of them. The question will likely be reworded, of course. If they are in a different field or a different seniority, they might tune the question. But always have answers to your own questions ready for when someone asks you what you just asked them. To help you build answers to your questions for yourself (where applicable), it will help to reflect on why you chose the questions. See Activity 4.6 to work through preparing answers for your own questions.

Activity 4.6: Preparing Answers for Your Own Questions and Fine Tuning Your Questions (Repeat When Your Questions Evolve for Different Events or When You Change Your Aims)

Activity goal: Build responses to your own questions written in Activity 4.2.

Purpose and benefit: Have some refined answers worked out beforehand that you are comfortable with, so you are less likely to get stuck without something to say.

Activity steps:

For this activity, I want you to take a moment to reflect on why you chose your seven questions for networking events (Activity 4.2 above). This is not a test but just a step to help you improve your questions further.

1. Take each question and ask, how would you answer? Would those answers be interesting to someone else at a networking event? Why/why not?

2. On consideration, do you want to fine tune these questions? If so, why? What is it about the first set of questions that made them robust or made some of them need updating? What does that tell you about what you are learning about networking?

3. Update your questions if needed and keep in mind the insights you have gained as to what makes a good question.

Takeaway: People will often turn your question back to you very naturally, "and what about you?". Alternatively, you can add to the conversation by volunteering your equivalent answer to what you asked them after they conclude speaking (where appropriate). For example, if you ask, "what brought you here and have you been before?", it is polite for you to share that information too in response. Ultimately you are practicing conversational skills that you will be able to use in the workplace too.

For Those of You Still Thinking Networking Is All Too Much

I have one more technique you can use. This is for those of you who, within a culturally and socially appropriate setting, still cannot think about talking to a stranger by choice. I want you to try practicing on strangers that have no impact on you personally or on your career. I want you to try to interact with them in a way that stretches your comfort level a little. A good choice of person is a type of person who you see on a regular basis who you would not normally interact with because of your tendency not to interact with those you do not know well, but that others would interact with. One example I use is the person at the supermarket or other store checkout. In many countries, the interaction with someone at the checkout can be anything from nothing to an extended conversation that actually slows people behind you down. While I do not want you to slow everyone down, can you practice increasing your interaction with someone like this? If you would normally avoid eye contact where others would make eye contact, try looking at them and smiling. In the next stage, can you increase this interaction with a "Hello, how are you?" and then politely answer their question if they return the "How are you?". This increase in interaction might take a while, but there is genuinely nothing to lose. If you would normally do that already, then can you ask them how their day has been or if it has been busy/quiet *etc.*? If you are worried about freezing up, have your question planned beforehand.

This specific example might not work for you. You might try the librarian, the bus driver, someone who serves your food or just someone waiting by you for some reason. The idea is just to practice in a way that you can control, that helps you learn to get used to talking to strangers.

You will likely still feel the spotlight effect and think that everyone is watching you. This is excellent practice for reminding yourself that the spotlight effect is really just a trick of our brains. It is very sensible for our brains to have evolved to highlight every little possible danger. But we do not currently live in the same world. There are rarely dangers around every corner. So, practice reminding yourself that all eyes are not on you. What were you doing last time someone talked to the person serving them when you were in line behind them? I am confident you probably didn't notice or were not sitting in judgement of what they chose to speak about and how they spoke. Hopefully, observing other people and how they act and react can help reassure you that very few, if any, people are judging you in everyday activities either.

How to Network Across Cultures

We first met the networking superpower of observation when we talked about the information one can gain just from watching how everyone else interacts with each other in the section 'How to Spot Who to Talk to and Who Not to Talk to'. In the example in that section, I encouraged you to try this technique to work out who you would most like to talk to, given their demeanour and how they seemed to be interacting with others. Watching the room is truly a superpower, though, and it can also be harnessed to understand cultural norms in the context.

There are two types of culture that we might refer to in the context of networking. Sometimes we just mean the culture of the business/discipline/ industry. One industry might be known for having a casual approach generally or about a specific aspect of work. For example, often people in the USA associate computer coders as having ultra-relaxed work environments. Other cultural work issues are entirely to do with the country you are working in (as represented in Figure 4.9), whether that is to do with the presence of alcohol at events, the level of mixing between people of different seniority levels or the actual format of the event (workplace karaoke, anyone?).

When I worked in a national laboratory in Colorado, USA, I was surprised at how casually people dressed for work. On the other hand, there was no shortening of names to nicknames or common short versions (such as Robert to Rob, Chainun to Nun). Shortening names is normal in Australia, and when I lived in Thailand, I'm not sure I ever knew my friends' full names as they used nicknames almost entirely. But in the USA, it happens to be considered rude to shorten someone's name unless they specifically insist you do so. So, these are examples of how we cannot simplify and make assumptions. But, with the superpower of observation, you will be able to note what is the norm in the environment you are in. A side note here: this does not just apply to networking, this is a superpower for any part of your life, work or private.

If you are working in a very different culture to what you have grown up in, or experienced while studying, I will assume here that you have put in quite a bit of time mastering the basic differences between the culture you came from and the one you are now working or studying in. You might feel like you are a beginner, but if you think about how much you have learnt, you are probably a star. There is so much to learn moving across cultures. Here we will deal with a few issues which are common in environments where people are socialising.

Preparation

Preparation is everything when developing cultural awareness, so you want to ask questions about the event (whether a formal or informal event), and you want to assume nothing. Read whatever information there is about the event, look for more information online and then

Figure 4.9 Working across cultures doesn't mean travelling. Each workplace has both its own culture and people from a wide range of backgrounds.

ask questions. Are there talks? If so, what will they be like? Who are the senior people for the night and what do you know about them? What do you know about the group organising the session? Have you been before and what was it like? Was there entertainment? Was it noisy or quiet? What people usually go? What do people usually wear at these types of events? Will there be food or drink? Will people go somewhere later after the event? When searching online for general information, please remember that even seemingly similar cultures can have differences, so be sure to look for search results around the specific culture that you are working in or across (for example, some people have made the mistake of assuming Canadians will follow all American customs or that people from the same continent are very similar). There are a large number of resources from the USA about other cultures, so it is very tempting to rely solely on those. But they might be quite unsuitable depending on where you yourself are coming from.

It is perfectly OK to ask colleagues or friends the above questions, specifically mentioning you are doing so because of the difference in work culture from what you are used to, whatever the type of cultural difference is (*e.g.* moving from sales to research or moving from the UK to China). For example, I know that in some disciplines it is normal for students to do a heap of networking and for formal events to be financially supported by the companies that the students will often end up working with. With other disciplines, the experience is the opposite. So even in the same city and coming from the same university, two students can have a very different perception of networking, plus they could have very different levels of experience and training for networking. Sharing why you are asking is much more likely to get you the information you need.

If the background is that you have not been living in that country's culture long and you want to learn more about how they network, just phrase it like that with a friendly colleague, fellow student or academic staff member and you will learn a lot. People are generally very keen to help others learn about their culture. This again taps into the idea that people like to talk about themselves or how much they know (in this case, their knowledge about their culture) and that many people like to help others. Even a very timid person will likely want to help you prepare. It is better if you have specific questions, though. People often don't understand how their culture (or their discipline or country) is different or similar to another. So, they may not know what to tell you that is helpful until you ask a specific question.

What About Online Networking?

While catching up with colleagues or classmates online has been possible for many years, since the COVID-19 epidemic, online networking has become very common. At certain times, this has, of course, been the only way many of us could catch up with colleagues. Online networking *en masse* has demonstrated that we can regularly hold meetings, conferences, conference calls, *etc.* with people from all over a state, country or even the world. Some people found they disliked online interactions,

Figure 4.10   Online networking can be incredibly useful.

while others, like this student in Figure 4.10, found that networking and developing relationships online was an enjoyable and rewarding process. Each to their own, but both can be useful.

Many larger online events have formal networking opportunities. They may or may not be labelled as such, but there is usually a deliberate chance to meet and talk to people. For example, a session might have breakout rooms to discuss a point that the speaker chooses to put to the attendees. While this is essentially a learning activity, it is also a chance to talk to people in your group in a small and likely friendly environment. There is generally a chance to talk to the people in the group quite casually. The conversation could flow to anything from something that happened at the same meeting, a topic related to the one set for the discussion group, or the random cat that just jumped on someone's keyboard.

By being fully present in these moments (*i.e.*, sharing your screen and your thoughts where you have them), you can learn a lot from the other interactions. And this is an environment many people feel more comfortable in, knowing that they can leave at any time if they need to. And let's face it. You don't even have to bid goodbye politely. If you really need to, you can just hop off and everyone will assume it is a bad connection. Not that I encourage disconnecting without saying you are leaving.

Online Networking Is Still Real Networking

It is really important to understand that the online environment still tells you a lot about the people and how they interact with each other. Who is chatting to whom in the chat? Who has their screen on and is fully engaged? Who asks great questions? Are there great questions you can note down and use in another context? Who are the attendees, what are their qualifications, and who do they work with? In an online environment, people come labelled! This is so useful for someone new to the group who does not know who everyone is. Being introduced to a middle-aged guy in a suit at the start of the night and not being able to work out which middle-aged guy in a suit he is by the end of the night in order to say goodbye, can be a bit embarrassing. But there is no risk something like that will happen if everyone is labelled on your screen. Remember to still do your research beforehand, but for online events, you can also do research on those attending during the session. This is especially useful if you did not know much about most of the attendees. You can use quiet moments to look up who people are, where they work, *etc.* Please remember to still pay attention to the speaker, though! There might be questions as online speakers work to include audience participation *via* polls or just open discussion.

Online Opportunities

Really importantly, some groups are continuing to meet up online, especially where people from that group are geographically dispersed. So, in huge cities like London, Tokyo, New York, Mumbai, *etc.* there is enough of a critical mass of most professionals to have a local discipline-based group, for example, an IT security professionals of New York meeting, or a Microbiologists of Mumbai meeting. For those with very niche interests or those in less populated areas, online meetings will continue to be very important, and continue to be a place where you can practice your networking skills while wearing your most comfortable slacks/trousers and shoes. For those of you who have internet access issues, this is not necessarily helpful, and I encourage you to look at participating from your workplace or university. If data is an issue, everyone will understand if you do not share your video. They will also likely understand (or even offer themselves) if you explain that your connection cannot handle data and ask if it would it be OK if everyone in the breakout session tried turning their video off. I honestly don't mind having to do this. It is not ideal but I, and most others, understand. After all, we have been in that position too.

Of course, there are some things that are missed online. It is hard to pull someone aside and have a quiet chat. The private chat function can fill that role somewhat, but it is not the same and it is often used by people who know each other already. However, you can use it to introduce yourself to a new person with a similar interest and suggest following up with an email or separate conversation (for example,

"Would it be OK if I reached out to you later by email or LinkedIn to talk more about ___? My email is___"). So, try it out and look for advantages in the online networking format and see what works for you.

How to Find Networking Events

First, I need to emphasise that, as covered in 'Why and When to Network' near the start of the chapter, there is no one type of networking event. The casual chat you see in Figure 4.11 is no more or less valuable than the conversations in Figures 4.1 and 4.3. Think of networking as what I mentioned earlier: a chance to get to know people in your industry or discipline. This can happen at a birthday or end-of-year party, a committee meeting, when you log in early for an online call, or at one of the larger events that people most often associate networking with. One of the best networking occasions in many environments is just asking someone to join you for a tea, coffee, or snack. Of course, it is not appropriate or possible to ask everyone to join you for such a casual talk. This is where more formal and larger events come in. At a more formal event, there will commonly be people who are much more senior to you that you do not get to see most days. There can also be people from different parts of the industry, company, country, or world. Different types of events will give you different opportunities.

Figure 4.11 There are all types of networking events: running club, book club, professional organisation committees, regular staff meet ups in a café, annual meetings, career nights, volunteering, training sessions, *etc.*

How to find events that suit you can be difficult, but with some persistence and help you will work it out. When you hear of an event, ask the following about it:

1. Is it a regular event; does it repeat?
2. Are there speakers?
3. What is the attendance number like? (You might prefer a certain size.)
4. What is the tone of the event? (Some events are more friendly or relaxed, or they are for more junior or more senior people.)

Whatever questions you think of, ask them, and then make some notes. Even if the event has been held recently, there is a good chance that the event repeats. As a professional or soon to be professional, it is useful to know about recurring networking events in your industry or discipline in general, not just to work out what is on in the next month or so. Everything you learn can be tucked away and will become useful in due time. Extending this point, it can be very helpful to ask colleagues, classmates, careers officers at the university, or the academics that teach/taught you what events they suggest. And if you are comfortable to, ask why they suggest these events. What is it they are hoping you will be able to get out of the event? Maybe even ask them for tips and pointers about networking at that exact event if you are comfortable. As I have pointed out many times, people like to help others and they like to share their knowledge about their industry or discipline. Prior to asking questions, do not forget to use your best professionalism skills to find a time that appears convenient and then check with them if they have time for a quick chat. It is likely only going to take a few minutes.

A really good way to meet people and learn about your industry/discipline is to join a group associated with it. If you are a student, you can usually find a group like this at your university. Nervous about joining? Find a couple of classmates to go with you. Then you will get to know new classmates too. More networking! If you do not want to just walk up to someone and ask them to go, you can post something to a group chat or in class. In a big class, there will be other students that are interested in going along but do not want to go by themselves. You might also be introducing the group or event to some students that did not know about it. You are now immediately more informed than they are and can help them understand what the group is about. Or you can go and learn together. All of this is great networking with your peers.

Outside of university, many professional groups also welcome student members for a small fee, or sometimes no fee. They also have student rates for catered events, which are often very affordable. The type of events that a discipline or industry group holds will vary dramatically based on traditions, size and the general activity level of those organising the events. Activity 4.7 can help you work out what events you might attend.

Activity 4.7: Generate Some Search Terms You Can Use to Find Relevant Networking Groups/Events

Activity goal: Find relevant networking events to attend.

Purpose and benefit: When you first start, it can be difficult to identify networking events, as there is no published list or even a set type of event that makes a good networking event for everyone. Use these steps to get started finding events for you to attend.

Activity steps:

To generate some ideas for how to find suitable networking groups and events for you, try the following steps:

1. What are four words/phrases that you can use to describe your industry? For someone who is a microbiologist working in a food testing laboratory, their four phrases might be:
   - food testing
   - industrial biotechnology
   - microbial safety
   - food safety.

   The search terms you identify should apply to an industry but not specifically to your role or the role you hope to get.

   For example, there are lots of types of food testing. If you look only for bacterial food testing events, meetings, or groups, you might not find one. But if you look generally for food testing groups and events, there will be groups focused on food testing and food safety. Within those groups, some people will be interested specifically in microbiological testing of food in a safety context. Alternatively, there are a lot of people in the healthcare industries that care a lot about microbial safety and testing. Our theoretical young microbiologist might be interested in some of those groups despite not currently working in a healthcare environment.

   If you look only in your specific sub-discipline (e.g., microbiology), you may not find the people you are looking for. It might be too small an area, or people might be organised in a way you do not expect. For example, it is common to have a chemistry society. But biology societies are less common because people are spread between ecology, health, agriculture/food, genetics, microbiology, and so on.

Continued

Activity 4.7: Generate Some Search Terms You Can Use to Find Relevant Networking Groups/Events – (Continued)

If you are struggling with this task, reading more about the industry you work in, or those you are interested in working in, should help you to start to form some terms that will be useful.

2. Now, think about your training. What four words/phrases can you use to describe your discipline? Here, I'm going to use myself as an example when I was about three years past my bachelor's and yet to start my PhD. So, my discipline terms would look something like this:

- drug discovery
- pharmaceutical testing
- organic synthesis
- green chemistry.

I graduated as a chemist, but if I went to a meeting about drug discovery there would not just be chemists there. There would be doctors, molecular biologists, data scientists, computer modellers and much more. I would have been interested in conversations with all these people and they would be useful for me to hear from and to get to know. But if I had been looking for a meeting with the search term 'chemistry', I would never have found these other contacts. This applies to every discipline or speciality and is by no means unique to chemistry and biology.

So, remember that search terms that are too general or too specific will not give you useful results. Although terms that are very broad will give you some useful results, they will be hidden in the 3 million hits that are not useful. This is the same when looking for jobs to apply for (see Chapters 6–8).

3. Now that you have your eight words/phrases, when you complete an online search using a selection of these, do interesting groups come up that you can join or are there events that you might attend? If you are still having trouble finding anything, now is a good time to sit down with a classmate, colleague, friend, or family member to see if they can help you further develop your search terms. If you have struggled to get eight, then these people can also help you with new ideas. Additionally, if you are still a student or a very recent graduate, your university will usually have careers training that will help you with these tasks, or chances to sit down in a small group or one-on-one to talk directly to a careers professional. This is a great opportunity to build you skills in this space.

Note: Your search terms will evolve as your experience and career evolves. But for now, you have a good set to work on.

Takeaway: This technique will help you start to identify events to go to, or groups that might have events. The process will help you learn more about the local work environment too as you will be seeing names of professional groups and employers. This will add to your understanding of your industry or industries, which is always a good thing.

Student Peers as Part of Your Network

It is really valuable to understand that as a student, you have a lot of peers who will soon become people who work in different industries and in different positions, and that these future professionals can become an incredibly important part of your future network. While you are students, you don't really know where people are going to end up, and that isn't really that important. If you keep in contact with a lot of students that you studied with, over the years many will go into roles that are relevant in your network. For example, maybe you don't go into sales, but a few people from your lab class do. You may well want to purchase equipment or supplies in the future, or help a colleague do so. If you have someone you can trust in the sales industry, you are off to a great start. Even if their company can't supply something you need, as they are in sales, they are perfectly positioned to help you find the company you need.

Even if you are still a student, you can set up a LinkedIn or other social media account and start to connect with fellow students. As you approach graduation, this becomes increasingly important as some of those people will be lost to your circle very quickly after graduation. Finishing your degree is also a conversation starter when you are connecting with someone. Unless you know the person very well, it is polite to introduce why you would like to connect. This can be very simple, *e.g.,* "I'm messaging because it would be good to connect before we finish our degrees." or similar. Be sure to put in a small amount of time to maintain the relationship.

How to Work Out What to Wear

This can cause a lot of worry. We all want to look our best, and we do not want to be over- or under-dressed. Nothing aggravates the spotlight effect like knowing you are dressed very differently to everyone else in the room. However, I encourage you to understand that this is still part of the spotlight effect. People are not noticing you as much as you think. On average, in my experience, young people tend to both under- and over-dress. If you are putting a lot of thought into your appearance, you are more likely to overdress. Is this a huge problem? No, not really, at least not in most contexts. But let us think about it for those who want help.

So, if this worries you, where can you start? In some places there will be instructions for dress for formal events. In that case, it is important to find out what that description means for your industry and for your location. If you are not that lucky and do not have a dress description for the event, the best way to be informed is to ask. Talk to a colleague who might have gone to that event or a similar event in the past. Remember, people generally do not mind helping. Select a friendly and helpful colleague whose opinion you respect in matters of professionalism. If you are very new to the workplace and do not really have strong opinions on people yet, then maybe think about whether that person seems informed about how to dress nicely for the environment they are in, and as that environment changes. And again, remember that people generally want to help.

Asking Advice From Colleagues and STEM Professionals

What is better than one opinion from a colleague? Opinions from three colleagues. It is just so valuable to get advice from "the locals" that know about the culture of that workplace or the industry. Talk to more than one person to get different points of view. Remember that whether they are the same or a different gender to you might limit the amount they will be able to help you in their insights.

If you are a student, essentially the same suggestions as above apply, but it will likely be harder to find someone who has attended. You can look for photos or videos from past events or talk to a staff member in the careers office, ideally with experience in your discipline. Often there are student clubs that will help with events associated with the university and they will often have more experience with events. Remember, the careers office staff are often outside of your discipline and cultural norms differ with discipline. Often there are graduate students who are older that tend to have gone to a range of networking functions, so these students are another resource. Joining one of the university clubs will give you a greater chance of meeting older students that undergraduates often do not meet.

It Is OK If People Know You Are a Student

I often note that students worry about not having a professional wardrobe, therefore worry about not being able to dress like a professional. This is an important point. In most environments, it is very well understood that students are students and do not have professional wear. If you have a nice pair of clean and neat slacks/trousers or skirt, and a nice shirt (that often does not need a collar for women), you will fit in at most events in most places. The style and length of the garments, of course, changes from country to country. Important cultural norms like not showing elbows, shoulders or hair will be the easiest for you to pick up. If you are already following acceptable everyday dress in these cultural variations, networking does not add a lot more other than the possible element that people can be more formal. Will you look like a student? Possibly. But given that you *are* a student, that is totally fine in most environments.

Further to that point, students often worry about the expense of nice professional clothing. This can be a real issue depending on the price of clothing and shoes in your area and your financial position. In most cases, you can consider asking a friend or family member if you can borrow an item of clothing. In countries that commonly have second-hand shops, these can be a great resource for very cheap professional clothing.

A general note that applies universally: you will learn more, relax better and get more out of the event if you are comfortable. This applies to psychological and physical comfort (that is, what you are wearing), and you are the best judge of how you navigate that. You will also get much better at this with practice. So plan ahead but, if you happen to misstep a little, remind yourself of the spotlight effect and ask yourself how much you would care if someone made a slight misstep. The answer is probably not at all. So, take a deep breath and good luck with your networking.

Using Social Media to Develop and Maintain Connections

Professional Social Media Basics

People have a range of comfort levels with different social media and that's totally all right. Before we make any assumptions or decisions, however, let's look at a few key points that influence social media use for professional or career related purposes.

Some social media platforms are much better for professional interactions than others, but it might also depend on the industry you are in. Some industries also have a stronger presence in general on social media, while others have a presence that is more focused on a particular platform. So, if you are a forest ecologist *vs.* a financial mathematician, you might find people like you on two completely different platforms. Having said that, as your career evolves and you work in different areas and with different groups of people, your network will start to deepen and broaden as you work in diverse teams, so you are not just looking to network with people with the same discipline background as you.

The main social media platforms that can be appropriate for professional promotion and development are LinkedIn, Facebook, Twitter, Instagram, and Snapchat. LinkedIn is the platform that is associated mainly with professional presence and is often called Facebook for work. While some posts on LinkedIn are more personal than others, there is an expectation that the tone is professional. Of course, that does not always remain the case. You can moderate your feed somewhat by unfollowing people or groups that tend to generate unwanted content on your feed, but sometimes it is just as simple as not reading the negative comments. However, this is the case with all social media. If you join and connect only in the professional space, your feed will remain much more professional than if you mix personal with professional. So, it can be useful to choose one social media platform to be personal and another to be professional. You can have a personal and a professional account on the same platform, but this can get confusing. How are colleagues meant to know which profile to add? Are you going to use two different names? What if too many cousins, aunts and uncles turn up in your professional feed?

How do you like the impression that the profile in Figure 4.12 gives just on first impressions? It has a casual but not unprofessional feel to it. So, this is fine for a casual environment. Make sure you think about your environment when building your profile.

> *"Attend your different discipline expos, [the] main thing is to network and talk to different types of people: networking. Networking is the biggest thing and actually connect with them on LinkedIn, keep your LinkedIn profile up to date."*
>
> – Ojasvi, chemistry and physiology graduate and medical research assistant.

Figure 4.12 A professional social media presence can help you connect to people, but most importantly it can help you stay connected to those you meet elsewhere.

A social media account gives you a way to connect with others in your field if they are also on that same platform. With a billion users, LinkedIn is the biggest, which is one reason LinkedIn is popular. If someone is using social media, there is a good chance one of their accounts is with LinkedIn. Whichever platform you are on, though, the idea is that this is another important layer to navigating networking and keeping all your connections in one place. There is much less value in meeting a whole lot of people in-person and then not capturing that in the longer term.

Building Your Connections

Once you have a professional social media account, it is important to put a little time into building it and then maintaining it. The first step is to put in connections. Connections are usually to individuals but can be organisations too. A good place to start is with current or former classmates. If you have professional work experience, also send invitations to people you have worked with that you can find on the platform.

Be sure to send a polite email about how it would be great to connect on the platform. Sending a blank invitation is generally thought to be rude unless you have already told the person you will look them up online, or it is someone you know well. Sending a message with the invitation also allows you to remind people who you are if you don't know them well. You don't usually have a lot of words, but it can help you "sell" yourself a little. Maybe introduce what you are passionate about, a shared interest that you have observed, or mention that you have a few shared acquaintances (by name) so that they understand a bit better where you fit in within their network. The approach will depend on your comfort level interacting online, the relationship you have with the person you are approaching, and any issues of status or seniority. It is much easier to approach people that are peers or close to being peers. However, remembering all the reasons that people want to talk to people that are more junior to them, those cases still apply to online interactions.

"Networking is the cornerstone of all your options."

– Dylan, chemistry and psychology graduate, superannuation consultant.

On some platforms, the use of a computer instead of the app on a mobile device can make setting up these connections a bit easier. For instance, on one platform when you go to invite someone to be a connection, the "invite" button automatically sends an invitation on the app and there is not an obvious chance to write a message. So be conscious that some functions might vary from mobile device to computer and explore what works best for how you want to use your chosen platform(s).

Who Is an Appropriate Connection?

When you are thinking of the list of people to connect to, remember that those you add do not have to be very close work colleagues. Maybe there is someone that you come across every now and then that you enjoy talking with, but you don't see them regularly at work. That is still a good person to connect to. You can also ask your contacts to introduce you to people on the platform. Maybe one of your contacts often shares their colleagues' work on social media and you are interested. You can message the person to let them know you are interested (depending on the settings of your platform and account type), or you could get the mutual contact to introduce you to them. That would be a great example of online networking!

Remember that you will have additional connections you can make if you have done an internship or a research project with a group. Or maybe you have volunteered. When you connect to people on a given platform, you do want to be conscious that connecting with a broader range of people with tell the algorithms something specific about your profile and the feed you will see will be adjusted.

Your contacts control the feed you see, so you can quickly start to learn more about what is going on in your professional community (this applies to students too) as you add connections. This will include events and training (often free, *e.g.*, webinars and online learning) that are being promoted in your community. Pay attention to your feed and think about if there are gaps in the spread of information you are seeing *vs.* what you might want. You can think about how you might balance your network to fill these gaps or go to another source to fill these gaps. See Activity 4.8 to work through how to make connections on social media.

Connecting With Groups

Most organisations have a social media presence now, so you can also connect to organisations. It is a good idea to start with a few professional bodies in your region and then follow a handful of employers that you have identified that you might want to work with or at least want to learn more about. By putting these companies on your feed, you will see what they post each week. This can tie in very well with increasing your commercial awareness as a source of an organisation's background information *e.g.*, new products, strategic plans, partnerships, advertisement of projects or new products, funding success, *etc.*, which are all discussed

Activity 4.8: Developing Connections on Social Media

Activity goal: Identify contacts and connect to them.

Purpose and benefit: Before you can connect to people online you need to identify who you can approach (who do you know and who can you introduce yourself to) and how. By building your online connections, you will be better able to keep in contact with people, develop a news feed that can help you learn, and develop new contacts to help you navigate your career.

Activity steps:

1. Fill out a basic social media profile with a nice professional picture (informal is fine but save the group photos or swimwear photos for your personal social media).

2. Write a list of 15 people you know from classes and try to find them on LinkedIn. If you can't find someone but you have found a mutual acquaintance, reach out to them, and ask them if they know what that person is doing or if they have them as a connection. Remember to send a proper written invitation note mentioning why you want to connect instead of just pressing the invite button.

3. Do the same for work colleagues, or those that you have done projects or internships with. If you have a longer work history, then try to look for 25 connections.

4. Now look at the connections your current connections have (most platforms allow this, but privacy settings vary from platform to platform and account types). Can you add anyone from their connections? Maybe a colleague that you were not close to, but it would still be reasonable to connect to?

5. Add at least three discipline-related organisations and three companies that you want to follow. You can also include individuals here that you find inspirational who you might follow rather than connect to. For example, well-known journalists and authors, high-ranking individuals in your specialty, well known entrepreneurs, and anyone else who you think has important things to say.

6. Set a reminder to hop on your new social media account every 2–3 days to read the feed and think about who else you could connect to. Interact with those connections you already have by liking posts, commenting (e.g., "excellent work" or "interesting post") and sharing posts you like. If this feels like too much just "like" a few posts the first week and save commenting for the second week.

Takeaway: By the end you will have a functional professional social media account and you will have taken some beginner steps towards using it effectively. Capturing your networking connections in a social media account is a really good step in developing good networks. There are plenty of free training videos that can help you learn more about social media and extend you from this point.

in Chapter 3: Commercial Awareness. If you follow a few big organisations in a particular field, then you will start to get an overview of the whole industry. The above remains true no matter what the industry. If you are interested in learning more about charities or government organisations that you might work for, then you can do the same. You can also join alumni groups, which can help you find former classmates or work colleagues.

Won't People Know I'm Looking at Their Profiles?

When you look at someone's profile on social media, they will often be alerted to that fact. Don't see this as a bad thing. Professionals know that there are many reasons people click on someone's profile; it is not like you are spying on them. For instance, the searcher might be trying to remember where you work. Or looking for a photo because they met a few people recently and they are trying to remember which person is which. Unless you have a distinctive name, people will also click on you looking for someone with the same name, not even looking for you. I see it as a good thing if people look me up before I have a meeting with them. It shows they are paying attention and want to get all their information straight before we meet. This is a sign of an organised individual who wants to be informed, which I think is a great start to a new relationship.

How to Get Value Out of a Professional Social Media Account

Now that you have a social media account, you need to keep it active. This can be as simple as checking it on the bus on the way to class or work most mornings or during a regular work break. As you become comfortable commenting on others' posts, move on to sharing the ones that you think are important. Once you are comfortable with that, maybe comment on why you are re-sharing the post when you do. This is an important next step because you are adding some thoughts of your own. While it can be as simple as paraphrasing something important from the piece, it is an important next step in building your social media confidence and competence.

Your feed (and everyone else's) is controlled by complicated algorithms that look at your connection and activities before they show your post/comments to others. The more you post and the more interaction you get on those posts, the more your posts are going to stay on people's feeds, increasing the likelihood they get seen. No post, and no comments means you won't be seen at all. Each platform will have their own training videos, or you can look up people who talk about how to use social media to improve your skills. Importantly, it is all about what you are comfortable with, and different platforms and types of interaction suit different people. If all you are ready to do is read and think about what is in your feed, then start with that. You will learn a lot about those people and groups you have connected to and that is an excellent start.

Where to Next?

Now that you have read about how to prepare for networking, how to work out what events to target and how you can work out how to navigate networking for your personality, it's time to sit down and do some thinking and planning. A good start is to make a plan around a first event and then a longer-term plan around what you might do in the next 6–12 months. If you start to get overwhelmed because you are new to networking or it is causing you a large amount of stress, just look at one or two aspects to start with. It can also be good to start with a shorter or smaller event, or one that is in a very familiar environment. I strongly recommend starting at the question writing activity if you don't know where to start. However, it is important to understand that there is no wrong or right place to start. Everyone will come to this point in the book with a different amount of experience and with different levels of comfort when we socialise and build connections.

Networking is a two-way building of relationships. You will meet interesting people and learn new things; some of these will help you in your career and other times you will get the chance to help others. Importantly, it is a great way to hear about opportunities. Everyone can find their own ways to network, and you will all improve with practice using the strategies in this chapter, including reflecting or your successes and on where you need to build your capacity. I wish you the best of luck working out how networking can best work for you as you progress through your career.

References

1. N. Cowan, Working Memory Underpins Cognitive Development, Learning, and Education, *Educ. Psychol. Rev.*, 2014, **26**(2), 197–223.
2. T. Gilovich, V. H. Medvec and K. Savitsky, The spotlight effect in social judgment: An egocentric bias in estimates of the salience of one's own actions and appearance, *J. Pers. Soc. Psychol.*, 2000, **78**(2), 211–222.
3. C. Raypole, Extroverts, introverts and everything in between, 2022, https://www.healthline.com/health/extrovert-vs-introvert, updated on July 5, 2022, accessed on November 19, 2022.
4. R. Mendoza-Denton, The Spotlight Effect, in Psychology Today, 2012, https://www.psychologytoday.com/au/blog/are-we-born-racist/201206/the-spotlight-effect, accessed on July 27, 2024.

Portfolio Building

5

SOPHIE MCKENZIE

Deakin University, Australia

Understanding your professional self, including your relevant career goals, interests, experience, skills and what you'd like to achieve as a part of your career, is an ongoing and cumulative process you will revisit and revise throughout your life. You can progress your career through reflecting on your professional development activities (see Chapter 2: Transferable Skills and Reflection). Your time at university is a great time to work through this reflection and development process. You can develop a portfolio to document and reflect upon your progress in a semi-structured way.

A portfolio (sometimes called an electronic portfolio or ePortfolio) is a collection of your learning and assessments, skills development, professional experiences and extra-curricular activities and achievements. This will cover your time at university, but will also include other periods of work and life. This collection of documents (often called artifacts) is the "evidence" or records of your abilities and skills. These records help you keep track of your achievements and are priceless when planning and writing your resume/CV (*curriculum vitae*) and key selection criteria (KSC) responses (see Chapters 6 and 7 for more details on how to put together both documents). Figure 5.1 shows some artifacts you might like to add to your portfolio.

How are they priceless? Well, there are three ways:

1. Gathering these documents, reflecting on what you are capable of (see Chapter 2 to learn more about reflection) and what that means for the jobs you apply for, is very valuable professional development.
2. You will build a handy selection of achievements for when you want inspiration for putting together your CV for a particular role. It is amazing how easy it is to forget an achievement. Flicking through them is a great reminder that you have great things to include in your CV and other application documents.
3. You might show a potential employer some of the things in your portfolio. For example, a video communicating your work, an app in development, advanced drawings on specialty software, *etc.* You might

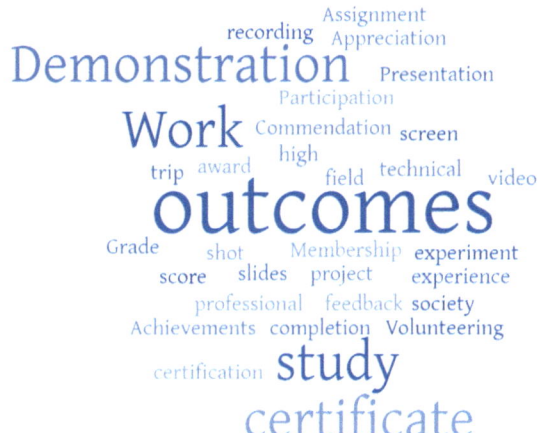

Figure 5.1   Artifacts in your career portfolio.

show them in an official setting like an interview or you might have work publicly visible that you can post or share (*e.g.*, on LinkedIn or on your own website).

Let's first look at the things that inform your career goals, or even lack thereof, before we look at how building a portfolio can help you refine and further understand your goals, strengths, and motivation for your career path. We will then look at what goes into a portfolio, why, and how to construct it. All the while keeping in mind the three uses for portfolios listed above.

Before You Build Your Career Portfolio, What Influences Your Career?

Students often come to university with a career interest or career aspiration based on their prior interactions with topics and professionals, plus their education history (both positive and negative). A career interest or aspiration is an early choice towards work activities and environments and can be developed through a variety of sources (see Figure 5.2). This interest can motivate study choice and signal early on in your learning journey the discipline in which a career may develop. Having an interest in certain careers is often where we start on our career journey, and while interest is a key predictor of career choice, there is a lot more that goes into informing the choices you make.

But not everyone comes to further study with a clear sense of what their future career may be, or even how they can make the most of their time

Figure 5.2   Sources of influence can come from a variety of sources and experiences; photo by Elijah Macleod on Unsplash.

at university to explore their career interest and build a description of their professional self (or career identity). Your career identity is your motivation, interests and competencies, and your understanding of how these relate to jobs/careers that are available to you. It takes time to understand your career interest thoroughly; reflection and building a career portfolio can help you with this process.

When students come to university, they will ideally participate in experiences that align with, and extend, their career interest. While students are all in different situations when studying (*e.g.*, pressure to get certain grades or to get into a certain career or need to work or help family while studying), it is important for anyone thinking about their career to consider what they want and how their skills and interests align with different careers and industries. After all, you are much more likely to succeed if you have a strong interest in the area you go into and/or your skills or tendencies line up well with the roles you take. To help you reflect on one of your key motivations, Activity 5.1 helps you explore why you chose your course or major.

Identifying the reasons for your study choices, as explored in Activity 5.1, can be very illuminating towards identifying your career interest. Having a career interest is key to career persistence. If you are interested in what you do, you'll find satisfaction in your work. IT Manager Mark Giles describes in the career snapshot (Box 5.1) how interest is very important when describing his career profile.

Engaging in various roles across your career can help you explore your career options and uncover what kind of work best suits you. We also

Activity 5.1: Why Did You Choose Your Course/Major?

Activity goal: Think about why you chose your course/major of study. An example of the relevance of your course/major to your career goals is described in the snapshot in Box 5.1.

Purpose and benefit: Understand why you chose your course/major to help you focus on your career interest and aspirations. Understanding your motivations makes it more likely you will be able to find positions that align well with your interests.

Activity steps:

1. In a few sentences, write down why you chose your current major. Consider in your response;
   - What do you like best about what you are studying? Why?
   - What don't you like? Why?
   - What aspects of personal motivation or experience from a prior role influenced the choice and why?
   - What type of job do you think you can get with your course/major? How/what informed this?

2. Using reflection (Chapter 2), examine what the points in step 1 mean and how your experiences since starting your course have changed (positively or negatively) your thinking.

3. Do you have any questions about whether you are in the right places? If so, write up to three questions that you can research yourself to further investigate these thoughts on a readily available platform (e.g., your internet browser or the careers office at your university).

4. Pick one of these and investigate it right now. What did you learn? How has that made you feel? Why? Did it raise any new questions to investigate?

5. Schedule in time to look at the other two question you raised in the next week or two (it should only take 10 minutes).

6. You should also plan time to come back to this exercise at the end of each trimester/semester. This is easily managed by putting a calendar reminder in now for a time after your exams.

Takeaway: Once you better understand your career interest, you can use that insight to build skills that relate to your interest. If this activity uncovered that you do not have a clear motivation behind your study choice, never fear, it may just mean you need to explore your career options further. Developing a career experience plan (Activity 5.4) may assist you to explore your career motivation further. Or you may want to consider your course plan and/or course choice. If this is the case, please consider talking to a course advisor at your institution.

> **Box 5.1   Working in IT Snapshot**
>
> Mark Giles is an experienced IT Manager. Mark graduated from IT in 1987 and joined the Australian customs service in the national capital of Australia as a technical programmer. After a couple of years, Mark moved to another major city to contribute his experience to another firm, with his expertise in mainframe computers being the main drawcard for the role transition. Mark commented that in these early years after graduation, he'd change organisations every two years in pursuit of a better organisation fit. "In the early part of your career, people do tend to change yearly or every two years, and quite often that's because they don't like the culture of where they are at." Learning about what you like, and what you don't like about work is important as it helps you understand what type of team you fit into and where you might like to go in the future. Leading from his work with mainframe computers, Mark completed further training with Sun Microsystems and Microsoft, out of Palo Alto (USA), to become a certified trainer and contribute to making computer software available to an international audience. This skills verification enabled Mark's career to progress to work for Queensland Rail to deliver a data warehouse project. At Queensland Rail, "I started doing strategy architecture work", which then led into other opportunities, "which sort of happened organically, more because I was following my interests, the work in strategy architecture that intrigued me and interested me", says Mark. "I've never had a formal career plan", Mark continues. After Queensland Rail, Mark joined SAP as an IT manager, finding the cultural change a huge shift in his professional working environment. This fast-paced, hierarchical structure "forced me to try and get to that level to try and be part of that group", describes Mark. In his role at SAP "learning new technology and processes became a constant in my work". "While the work environment was constantly busy, there was a common focus and approach to the work completed", states Mark.

must pair our interests and motivations with an understanding of what jobs are available where and why. Next, we will briefly look at understanding the availability of different roles. Associate Professor Sharon La Fontaine talks about the jobs she has had in her career, including where transferable skills can take you, in career snapshot (Box 5.2).

Understanding What Jobs Are Available

Having a realistic understanding of the jobs available (the labour market) is also important for anyone looking at applying for a job in the next couple of years. As Figure 5.3 shows, often we need to look for signs to inform us of what jobs are available. When developing an understanding of the labour market, it's important to look beyond assumptions of the type of

---

Box 5.2   Working in Science Snapshot

After completing a Bachelor of Science at Monash University, followed by an honours year in microbiology, then a PhD in the same area, Associate Professor Sharon La Fontaine completed a postdoctoral research position at Murdoch Children's Research Institute, working in "…human molecular genetics focusing on the role of metals in health and disease; the role of copper in a couple of genetically inherited diseases". After 4 years working as a research fellow, Sharon returned to the higher education setting, working in a US university research lab studying how plants metabolise and utilise metals, before returning to Australia to continue research on copper in health and disease. In 2011 she took up a teaching and research position, continuing with research and teaching infectious diseases and immunology, as well as professional practice, supporting students' development of their work readiness and career management skills. Now, a focus on career education forms a core part of her teaching practice, inspiring science students and graduates to focus on their career building and completing experiences to enact their career development plans.

When it comes to working in science, Sharon notes "it is very broad now". "From a biomedical science degree, we also see students branching into teaching, nursing, physiotherapy, dentistry, veterinary science", states Sharon. Students also take on roles in "sales, pharmaceutical sales, government roles, business development in pharmaceutical companies, in pathology, private pathology labs, consultancy companies, a whole range of different organisations". Adding to that, "some take on graduate roles in government, in a whole range of different government departments. It could be the Health Department, but we've had other graduates who have gone and worked in the security area. [...] Many will go on and do research and development type roles, requiring them to complete an honours or postgraduate degree before working as a research assistant". Studying science "opens many, many doors because of the teamwork, the independent work, problem solving and critical thinking", says Sharon.

---

people who work in a particular place or industry. A prime example is the assumption that students within science, technology, engineering, and mathematics (STEM) disciplines are high-achieving students who prefer technical roles, or even that they tend to stay in STEM focused roles. In contrast, previous studies suggest that students who study STEM related disciplines aspire to a variety of roles in and outside of the discipline.[2]

Some roles are STEM adjacent, and it is completely impossible to fill those roles without your STEM qualifications, *e.g.*, health, safety, environment and compliance, specialty consulting, STEM teaching and education support, some policy work, and specialty management. Others are science

Figure 5.3   Looking for signs in the job market; photo by Eric Prouzet on Unsplash.

adjacent, and you may well be chosen for your STEM skills, but people with other skills will also work in the same group or team, providing a well-rounded skill set. General consulting, management and policy work can all benefit from a strong STEM background. Governments generally want STEM knowledge in most or many offices, like environmental protection, health and safety, wildlife and parklands, education, disaster management, weather/climate, data management and cyber security. In these roles, you might not rely on your STEM discipline knowledge regularly, but you use your STEM degree's transferrable skills and still might be called on for STEM knowledge during, for example, special projects.

Jobs are often categorised into broad professions that relate to a field. At a high level, these professions are reported as labour market statistics, providing a high-level summary of what types of jobs people are doing. For example, when thinking about the professional field that encompasses Information Communication Technologies (ICT) we can look for data on jobs to give us a sense of who is generally employed in this area. Most positions now consider digital skills to be a required attribute; however, what is required to work in an ICT related job? If you look up jobs in ICT, you will find job titles such as ICT sales professional, ICT mangers, business and systems analysts, ICT trainers, software and applications programmers, multimedia specialists and web developers, and computing networking professionals. These roles all require a bachelor's degree qualification.

Just like ICT though, each area of science or engineering has a whole range of specialties. So, there are chemical, biological, and engineering sales roles, managers, consultants, analysts, data specialists, trainers, and much more. To help you find more information about careers in your area of interest, complete Activity 5.2.

Activity 5.2: Where Can I Get More Information About Jobs?

Activity goal: Better understanding jobs that are available in your local area.

Purpose and benefit: By understanding a broad range of jobs you can be informed when selecting a job role.

Activity steps:

1. Find the website for your local careers service and national professional society. On these websites you will find the information that relates to jobs and careers. These websites may also describe jobs available in your local area. If you are not sure where to start, look at the Organisation for Economic Co-operation and Development (OECD) https://www.oecd.org/employment-outlook/2023/. This annual report summarises global employment trends.

2. Using the information identified in the chosen website, note down 2 to 3 professional fields you could consider working within.

3. Once you have identified professional fields you would like to work in, find 2 or 3 jobs that are available in this field. For example, if you want to work in 'microbiology' you may find the job role of a 'lab technician'.

4. Using the job roles, you have found, note down the job requirements, including experience, qualifications and skills needed.

5. Think about what has surprised you. Is there a role you hadn't thought of? Are the skills lists too long? Keep in mind that advertisements are "wish-lists" so you should aim for meeting about 60–70% of the criteria and all the qualifications. We talk about this more in Chapter 6: Resumes and Cover Letters and Chapter 7: Addressing Key Selection Criteria and Interviews.

6. File these notes away for reference when you are looking for roles. You will come back to these jobs in a future activity.

Takeaway: By understanding your local career services, national government, and industry bodies, you can feel more connected to your local labour market. Through this review, you can understand if holding a degree in the labour market in your area is very common or whether a two year or technical (non-university) qualification may be enough.

Reasons to Build Your Career Portfolio

It can feel overwhelming when thinking about building your career (see Figure 5.4). Knowing what to include when preparing documents that describe your professional self (such as on your resume/CV), when to update them, and how to share with others can be daunting. In today's job

Figure 5.4   Thinking about your career can feel daunting at times but there are some simple ways to get started; photo by Luis Villasmil on Unsplash.

climate, often having an online profile or brand has become important for networking and getting a job. As mentioned in Chapter 4: Networking, LinkedIn is one of the most common platforms now for professionals to demonstrate their career and network with those in their desired job area. However, it can take time to build a successful description of your career to put on LinkedIn, with many students unsure of what to include. One of the best ways to work towards a description of your career that you are happy to share *via* technologies like LinkedIn is by collecting your outcomes and achievements (as discussed further in the next section) from your time at university into a portfolio.

As a collection of your learning, assessments, outcomes, and achievements from your time at university, your portfolio is like a file repository, at least to begin with, but it is what you do with your portfolio collection that matters. As mentioned in the introduction, a portfolio helps you: (1) organise your things, giving you a focus to reflect on experiences, what you learnt and how you grew from them; (2) have a set of materials to refer to when writing documents for job applications; or (3) have a collection of material you might like to share part of with potential employers. It really depends on the type of job you are applying for as to whether you include your full career portfolio as a part of your job application or whether you use parts throughout.

Organisation, Articulation and Reflection

You will find the preparation and reflection (see Chapter 2) you do in preparing your career portfolio will enable you to better articulate yourself in your resume and/or cover letter. Understanding and articulating your skills confidently and professionally is key to securing a job, as you can present yourself clearly during application and interview. It also makes interviews and the application process less stressful if you are better prepared and know what to say and when (see Chapter 6: Resumes and Cover Letters and Chapter 7: Addressing Key Selection Criteria and Interviews for details on being well prepared for the job application process).

Articulating your experience beyond grade outcomes is an important activity in the job search process, enabling your individual attributes to be demonstrated to prospective employers. This includes demonstrating your transferrable skills, as a career portfolio is not necessarily tailored towards a specific advertised job role, unlike a job application or resume that you will update to meet a specific advertised role. It has been shown for some time that building a career portfolio during your studies can have a significant positive impact upon your ability to present yourself to employers.[1]

> **Hint:** Your portfolio is also a space you may use for personal development while you are studying, enabling you to reflect upon your growth, relating this growth to your skills and expectations for your future.

In a portfolio, reflection can be used for self-evaluation; to present your learning; to seek feedback; or as a tool to explore and learn about your career. Reflection enables you to make an evaluative judgement of how your activities and outcomes demonstrate your capabilities.[3] Reflection also enables you to bring together the various artifacts collected during your time at university into a cohesive description of your professional self and your career goals. It can take time to reflect meaningfully on your outcomes and achievements during your time at university, and when you begin it may feel trivial. You may wonder why you need to describe what your outcomes and achievements mean. Over time and through practice, you will see that reflecting upon yourself and your skills provides useful insights, not just into what you have done, but how you will act in the future. Reflection enables you to have a greater understanding of your processes and the thoughts behind them, promoting 'thinking about thinking' and a deeper connection to the experiences and activities that define you.

> *"We learn from reflecting on our experience."*[4]

The process of reflection is cyclical, in that you will keep coming back and reflecting upon your portfolio artifacts, extending upon the way in which you select and present artifacts.[4] You will regularly revisit and reflect upon your portfolio collection. Next, we discuss how you can use the artifacts in your portfolio to reflect upon your skills development.

Communication skills are broadly considered one of the top skills required across most jobs, regardless of discipline. Regular polls of employers into what skills they seek in new graduates highlight communication skills, both oral and written, as essential. During your time at university, a portfolio can be a great way of demonstrating your ability to communicate as it enables you to bring together multiple artifacts through reflection to demonstrate your ability. It can be used with potential employers to demonstrate your capabilities, and the act of creating your portfolio will inform other professional documents, such as your resume. For example, when discussing how to communicate within a professional environment, a student may use outcomes from a group project and/or their placement experience to reflect upon their ability to communicate. The example that follows demonstrates how a reflection in your portfolio can outline your communication skills:

> *"During my university group project, and my 3-month placement, I utilised my oral and written communication skills in many situations. I discovered in my group project the need for clear communication of project objectives and progress, regularly updating my teams' members on my progress through updating the project management documentation and discussing my progress during team meetings. My approach to communication was validated during my placement through my supervisor and team commenting on how clearly I updated the team with detailed notes and logging. This showed me how important regular communication is for team cohesion and my own productivity. My focus on communication saw my placement outcomes exceed my supervisor's expectations."*

In Box 5.3, Associate Professor Cai Wilkinson describes how reflection can assist students in articulating what do to with their degree, which is relevant even for students in STEM in today's varied job market.

> **Box 5.3   Reflection to Assist Your Skills Showcase**
>
> Associate Professor Cai Wilkinson places a particular emphasis in her teaching on developing students' employability skills. Cai asks her students to consider, "How does this affect your understanding of the world? How do you want to be a part of it?". Reflecting upon her experiences from teaching in higher education, "across the board we need to be helping students in developing their communication skills, and while particular communication skills are going to have different flavours in the different disciplines, at the same time communication skills should be transferable as well". Cai argues that students should use reflection as a personal development process in preparing a career portfolio. She suggests that during their time at university students should, "explore their options for using their training and that technical knowledge potentially in other situations and in other circumstances", and should present the outcomes of this *via* a portfolio. The challenge during study for every student is "articulating what we do with a degree", argues Cai.

Collecting Your Documents for Later Reference

This is self-explanatory, although the importance of this step should not be underestimated just because it is simple. Simply collecting your achievements in one place over time means nothing is forgotten. You have a treasure trove of information to draw from when you are writing a job application. For example, let's say you are asked to talk about your written communication, verbal communication, time management and attention to detail in an entry job application. For details of how to do this, go to Chapter 6: Resumes and Cover Letters and Chapter 7: Addressing Key Selection Criteria and Interviews. But for now, think about the fact that you will need some examples to think about. Ideally, all examples will be different, and your examples will highlight your strengths while not highlighting your weaknesses. You also want to tune it somewhat to the type of role (*e.g.*, including something for coding in a coding role, including a lab project for a lab technician role, *etc.*). It's a good thing you have a big folder of your achievements and experiences to refer to in order to find the best examples to write about!

> **Hint:** Remember to revisit your portfolio regularly to add and reflect. We suggest setting up a regular calendar appointment for an hour every month or so (depending on how much collating and reflecting you have to do).

Having a prepared and clear description of your professional self will help you be best prepared for employment and enable you to stand out from the crowd. Having a portfolio collection also allows you to make different descriptions (or presentations) of your portfolio collection, curating the resources you have collected towards a certain goal, such as preparing a showcase of your teamwork skills, or preparing a summary of your career. Some universities provide portfolio tools within the learning management system, enabling students to add their learning, assessments, outcomes, and achievements to their portfolio at any time. If your university does not have a portfolio system available, saving your output to your local computer storage, or even a personal online drive, is also a suitable approach to collecting your outcomes from your studies. The portfolio is much more a concept, not a set way to store files.

> *Hint:* Whatever way you choose to use portfolios in your life, you will draw upon your experience from your time at university through an interactive, engaging, and individual account of your professional self. Preparing and presenting a portfolio is a varied activity for everyone and no two portfolios will look the same. Through the act of collecting your outcomes and achievements in a portfolio, you will deepen your thinking about your experiences, developing your learning and professional identity.

Dr Coleman, an art educator who uses portfolios all the time in their practice, argues in career snapshot (Box 5.4) that portfolios can be a great way of developing your unique brand through developing a story around your professional journey.

What Do I Include in My Career Portfolio?

When thinking about a possible career, many factors will influence your ideas of what a career is and can be and ultimately what you put into your career portfolio. Both informal and formal things can influence your career. For example, you will build your career perceptions through interactions with your family, friends, and people. Your social network, or your social relationships, are often a key source of career related influence. Your prior experience with those working in the field, with those unemployed, or yourself working as a professional (in any capacity) can also influence your understanding of a career, as does your academic achievement related to that discipline (*i.e.*, you may be more interested in a certain job based on the studies you did that relate to that

---

Box 5.4 A Designer's Journey Snapshot

"I am a visual artist, but also love technology", says Dr Kathryn Coleman, an Art and Design Educator from the University of Melbourne. "My dad is an accountant, but was also super into tech, so I was trying to find a way, particularly through school, and then at University, to bring these things together" as a profession. Through undertaking a Bachelor of Art Education Degree at the University of New South Wales, Kathryn was able to bring together her love for technology into a professional career, educating art and design students in high schools and now teachers.

Working in an interdisciplinary area makes finding a career path both a little easier and a little more difficult, describes Kathryn. "For a student who is in art and design, the disciplines are already interdisciplinary because they're not core professions, students will end up working across disciplines." So, when developing a career profile in an interdisciplinary area, "the profile is how to actually engage in multiple spaces and places". Current secondary and tertiary education pathways can result in students learning about "something that they are good at, rather than potentially having the chance to think about what interdisciplinary might look like or feel like", says Kathryn. "It's far easier for a student to say, 'I'm a designer, I'm going to study architecture, and then I'm going to become an architect or a visual communication designer', or 'I'm an artist and I go to art school, and then I major in something'", taking forward a more linear career path.

Kathryn uses portfolios as a tool to work across disciplines and evidence a career profile, arguing that portfolios are not just the result of "working in a studio". Portfolios these days are often electronic, enabling you to develop your career profile to be disseminated online. A portfolio enables you to consolidate your skills, but also allows rich experiences to be articulated. In a career portfolio, Kathryn says that "you have to be able to construct the story and reconstruct that story all the time, to tell the story of who are you" throughout your life.

---

job area). Experiences that help build your employment skills and ability can be developed both inside and outside of your schooling. Engaging in experiences that support your professional development more broadly is also valuable in building your understanding of what a career looks like. In addition, these experiences are the thing you can demonstrate and reflect upon in your portfolio. Career experiences can take a variety of forms. They can be a small moment in a day (such as reflecting on your achievement in your latest assignment) or can encompass months of more formal work (such as taking on an internship or placement). Table 5.1 describes a variety of experiences that can contribute to building your career and can be things you reflect on in your portfolio.

Table 5.1   Career building experiences and how to demonstrate them in your portfolio.

What is the experience?	How would you demonstrate this experience in your portfolio?
Completing a **work experience/placement/internship** (in your degree or outside it) to develop experience working in a professional environment.	In your portfolio, this would include a description of work experience outcomes and achievements from your time in this professional situation. You can make a focus on the skills you developed through this work experience, putting a focus on your professionalism. You may have a letter of thanks or recommendation from your work experience supervisor.
Volunteering with a local club or community organisation. This could be your local sporting or hobby club or could be volunteering at your local food shelter. Or you could volunteer at your university.	In your portfolio, this could include a description of your role when volunteering. You can make a focus on the different people you worked with, and the networks built through volunteering. Volunteering can also help you evidence your ability to take on a leadership role. For your efforts you may have been presented with a certificate showing your commitment through volunteering. This could be included in your portfolio.
Complete an **exchange or study abroad program** as a part your course. Many universities have partner institutions to formally support student exchange and provide you with credit for the study you completed at a partner university.	In your portfolio, this could be a photograph and a description of your experience, including a demonstration of project work or study achieved while on exchange. You can describe your experiences with other cultures and demonstrate your broadened perspective on global issues.
Engage in **peer mentoring.** Peer mentoring is when you work with other students to support them during their time at university.	In your portfolio, this could include a certificate of your participation as a mentor, along with description of how the experience built your interpersonal skills.For example, you may assist other students in planning study routines, finding academic resources, organising themselves, or you might provide more general advice on navigating university.

Table 5.1 *(Continued)*

What is the experience?	How would you demonstrate this experience in your portfolio?
Join a **social club** at your university or in your local community. There are often many different clubs available on different interests and hobbies, such as sports or film.	In your portfolio, you could demonstrate your participation in these clubs through a description of activities completed with the clubs. You may be able to participate in a variety of different roles as a part of this club, which provide opportunities for you to develop your leadership and interpersonal skills. For your efforts, you may have been presented with a certificate showing your commitment through the social club. This could be included in your portfolio.
Participate in **competitions or challenges,** such as a hackathon. These are often available online for anyone around the world to participate.	In your portfolio, you could demonstrate what you learnt in the competition or challenges through a description of the event and the resultant output. You may describe how the competition improved your transferable skills, and ability to work with others.
Engage with the **career services** at your university or in your local area if one is not available at your university. Often these services are free for students and even graduates of the university	Visiting a career service helps you to have a better understanding of yourself, which is helpful when putting together your career portfolio and any job application. In your portfolio, you can use the outcomes from visiting a careers service within your resume/CV and in the introductory section of your career portfolio (see the section 'Setting Up and Sharing Your Career Portfolio').
Join a technical **professional society** or industry body. These societies help new and established professional connect and build their professional networks. Often these societies have discounted or free memberships for students.	In your portfolio, you may demonstrate your membership to a professional society. They can provide you with a certificate of membership to include in your portfolio. By joining you will be able to network and engage with professionals already working in the area (see Chapter 4 for details on how to plan and navigate networking).

Box 5.5 describes how work experience or a placement can form an important career experience.

There are many ways in which you can demonstrate your career experience in your portfolio. Maybe you attended a career event and talked to some employers? Did you attend a guest lecture on a special interest topic? These examples show the engagement you have with professional

---

> ## Box 5.5   Work Experience/Placement as a Career Experience Snapshot
>
> Associate Professor Cai Wilkinson suggests a good way to explore career potential is through work experience, such as an internship or placement. Cai advises to "think of it as an opportunity to explore and to find out what really drives you. What really motivates you and remember that there's no such thing as a failed experience". An internship or placement can be completed from various professional situations, from working with large or small teams. There are lots of different opportunities. Sometimes the team will be very much in your specialty, but working with a team from a different background or a diverse background can also be a very rewarding experience. "You've got a chance to really show your skills, and also to get a wider picture, if you are willing to look beyond the prestige opportunities. [...] Be willing to give things a go. But also, be willing to ask for help along the way. It's not all on you to get it right from the start."

activities or networking events, and this can be described in your career portfolio. Activity 5.3 will get you started on collecting artifacts for your career portfolio.

This list of career building experiences that influence your career is not exhaustive; there may be many more ways in which you can expand your ideas of a career. Examples of other career supporting experiences may include academic activities such as lab/fieldwork, a coding competition, an essay, or a group project. Part-time work or vacation programs are also examples of career experiences.

Linking Your Schooling to a Career in Your Portfolio

It's also important to consider how your formal learning and assessment from your discipline studies influence your career experiences and your portfolio. For example, your assignments, submissions such as your programming code, a lab book, or poster presentation are all things you can use in your portfolio to demonstrate your career interest and experience. Your studies can also help you evidence your non-discipline skills development, such as outcomes from group projects, for example, a product developed for a client. You will show the skills you demonstrated to achieve the project (*e.g.* teamwork, communication and interpersonal skills). The assignment examples will show your discipline, such as your technical abilities, while also showing your broader transferrable skills put in place while applying discipline knowledge (*e.g.*, research, organisation, time management, and writing skills).

Activity 5.3: Collecting Artifacts to Reflect On in Your Portfolio

Activity goal: This activity will add useful artifacts to your portfolio collection so that when you build your portfolio and reflect on your achievements, you have evidence available to support how you describe yourself.

Purpose and benefit: By adding to your portfolio collection regularly you will soon build a comprehensive record of your achievements and outputs that you can reflect upon in your career portfolio.

Activity steps:

To complete this activity;

1. Consider what learning and assessment artifacts you currently have from your time at university. What artifacts on the above list do you have?

2. Collect at least three artifacts you have available and save them to your portfolio collection. You can start to curate these artifacts by simply saving them in a folder on your computer. Or, if your institution has a portfolio platform, save these artifacts into that portfolio platform. Often this portfolio collection is available on the institution platform after you graduate. Storing your portfolio collection online is also suitable, however, do ensure you store your collection in a secure environment and one that will enable you to continue to access your files once you finish university.

Takeaway: Having your artifacts saved in a relevant portfolio location will ensure you build a collection that considers many and diverse artifacts. With these artifacts you can then reflection upon your professional self.

Hint: In your portfolio, you may like to archive your academic transcript from your current and prior studies, so you have a record when needed. These transcripts are important documents to evidence your learning journey. You may want to reflect on your whole course, or course to date, and see what story it tells of your learning.

Note that many employers have no interest in seeing your transcript, but this is very dependent on the environment you work in. It is a good idea to look for people to ask about what is normal in your environment. This can be the career staff at your university. An internet search will help sometimes too, dependent on how niche your environment is. In many places, the marks on your resume are also not of interest. Although quickly noting a degree average in brackets near the course name is seen as acceptable in an environment that does not need your grades.

You may also have certificates of achievement, such as honours or awards in relation to your study. This could be a commendation on your study outcomes from a particular year of university, or it could be a communication award for an essay you wrote. Having a copy of these saved in your portfolio can be useful evidence to certify your skills development. While in most places it would not be suitable to submit certificates as part of your job application process, collating your achievements and having them as a reminder can help you when reflecting or writing the application. It is both important to remember your success and useful to build a little table of achievements in your resume for important and relevant achievements or additional qualifications. It is important to make this a relevant list and not too long. About 4–5 achievements/awards (*e.g.,* top student, team award, best essay, most helpful student, best prepared, *etc.*) and additional qualifications (*e.g.,* first aid, mental health, certificate in tutoring, *etc.*) is likely enough in most places. This sort of "catch all" place to put great additional achievements that don't otherwise have a home in your resume can be useful, but don't overdo it. And, as always, ask around to see what is normal in your environment.

While you are at university, it is a great time to engage in a variety of experiences to help you explore and expand your career. It is important you consider and identify a variety of activities and experiences, those that broaden both your academic skills and industry experiences. Experience, beyond study, can be large or small, yet all impact your understanding of your professional self and where you might like to work. If you are accepting of new experiences, your career decision-making will become more informed, you will have a clearer view of where you want to start in the workforce. And how to get there.

Building a Career Plan

A career plan can help you seek out experiences and activities that purposefully build your career. As described in the section 'Reasons to Build Your Career Portfolio', there are many formal and informal experiences you can engage in to build your career plan. If you haven't got a career plan already, then use Activity 5.4 to help you develop a timeline of things you will do during your time at university to engage in experiences that will build your career goals. This plan is a timeline that will map out your career experience, which will take place alongside your

Activity 5.4: Career Experience Plan

Activity goal: Develop a career experience plan that outlines a brief timeline of the activities and experiences you plan to undertake throughout the rest of your studies to help you explore your career. Experiences that influence your career are varied, with some examples described in Reasons to Build Your Career Portfolio (What Do I Include in My Career Portfolio?).

Purpose and benefit: By having a clear plan, you will be better able to work towards achieving that plan. Even if this is a draft and you redo this activity a couple of times while you work out what you want, it is the process of working out what you want that is the focus here.

Activity steps:

1. Fill out Table 5.2 to plan your career experiences. You can add more columns as necessary to add more career building experiences to your plan. For each career building experience, complete each column as described below.

2. Provide a brief title to describe the experience you plan to pursue. Look for experience both inside the curriculum (project work, optional internship, volunteering to be team lead, optional field work unit, *etc.*) or outside the curriculum (part time job, external internship, volunteering, tutoring, teaching yourself a skill, budget, or free training, *e.g.*, online learning).

3. Next, describe what knowledge/skill(s) you would develop in this experience. For example, are you hoping to develop discipline knowledge, hands on discipline skills or transferrable skills?

4. How would you develop these skills in this experience? For example, would you be giving oral presentations regularly (communication skills), managing a lot of data (organisational and data management skills), or building specialty equipment (hands on discipline skills)?

5. When will you do this experience? Consider when you will complete this experience. Will it be in the short-term or something towards the end of your studies? Write down the year and the teaching period (semester/trimester) when you plan to complete it. How long will it take? How will you balance other commitments? While it is true that it takes time to develop skills and experience, an hour twice a week over summer can really add something significant and need not interfere too much with other things you need to do.

6. Once you have done one experience, find someone to talk it through with out loud and ask them what they think. Explaining what you were planning and why can help improve any plan. This person doesn't have to

Continued

Activity 5.4: Career Experience Plan – *(Continued)*

be in the same discipline. Look for someone who's opinion you trust who knows something about STEM workplaces in your environment. Remember, talking to multiple people can add depth to how much you can improve a plan. Review your plan, keeping the feedback in mind.

7. Move on to planning your next experience. Keep in mind that ideally you are building your experience with a particular target, but it can be a general target, *e.g.*, better communications skills. Also keep in mind a spread of experiences and skills is useful. Most people don't need to plan many similar experiences.

Takeaway: Engaging in experiences that expand your career prospects can be managed by finding opportunities both within and outside your curriculum and can be built up over time.

---

formal studies. In addition, you will include important dates and/or deadlines for the selected experiences (*e.g.*, application deadlines for study abroad or exchange programs or vacation work, campus information sessions, *etc.*). If you are currently working while studying, consider what experience may complement your professional development. If you are not yet working, consider when you will plan for formal work experience opportunities, such as an opportunity for a placement or internship as discussed in 'Reasons to build your career portfolio'. Or think about what casual work you can take that will help develop important skills for your future career. You don't have to know exactly where you want to go. But knowing the industry or area is helpful. For example, if you think lab work is not for you but you have identified technical sales as of interest as you like working with people, can you get sales experience? Are you interested in consulting? Have you developed your customer service skills to help in consulting? If you are interested in roles with a leadership element, have you led a small project or team (independent of professional settings)?

Pitching Yourself to Start Your Career Portfolio

A great way to start your career portfolio is by capturing some of the skills and experience you have developed so far. A career summary is often used as the start of a resume as a way of highlighting your professional self and how you align with a job. In some environments, a

Table 5.2 Planning your career experiences.

	Career Building Experience 1:	Career Building Experience 2:	Career Building Experience 3:	Career Building Experience 4:
Brief title of experience/activity to be completed.				
What knowledge/skill(s) would you develop in this experience?				
How would you develop these skills in this experience?				
When will you do this experience?				

career summary will be valued more than others. This is because, unlike what you write in your resume, you often aren't backing up what you say with details and examples. If you can, include small points of detail. For example, instead of saying you are a self-starter or motivated and an independent learner, list the online courses you have done, the IT certificates you have added next to your degree, or the language you taught yourself.

Often, a career summary is targeted towards a specific job role. But a career summary can be adapted to suit other uses, such as a summary you place on LinkedIn, or as your 30 second 'elevator' pitch you use to introduce yourself when networking (see Figure 5.5). While these short statements are important, your career is much broader than a simple statement. It takes time to develop your career summary to be applicable for use in a variety of situations where you can 'pitch' yourself professionally to potential employers. Pitching yourself is about developing your personal brand and then taking that brand with you wherever you go as a way of showing your career and what you'd like to get out of it. While it may feel difficult to complete a career summary/pitch while you are still studying, practicing what you put in your summary and how you articulate yourself is a useful activity to begin from the first year of your studies. Your career is cumulative, so your summary will change frequently during your life. Typical summaries of first year students are given in the next example:

Figure 5.5 Pitching yourself and selling your skills; photo by Headway on Unsplash.

"An enthusiastic and mature first year Bachelor of Cyber Security student who has high levels of self-motivation, along with excellent customer service and problem-solving skills. Looking for a casual Junior IT Support position to gain practical experience in the field while studying."

"First year commerce student looking to Major in HR management. High levels of written/verbal communication skills and pays great attention to detail. Seeking an Internship within HR to help build skills. Enjoys building strong and positive relationships within the workplace. Career goal is becoming an experienced HR manager."

The career summary you create is just the starting point. Adding to it by reflecting on how the various professional development opportunities you have completed have complemented your career will see your summary change every year. Your education, experience, networking, success, failures and response to opportunity will update your career summary/pitch. Your pitch becomes a part of who you are, and you take it with you when you complete any professional engagement, particularly during a job interview. Pitching yourself first requires you to summarise your skills, experiences, qualifications, and interests, so think about how much you know already, as this can boost your confidence in preparing your pitch. A simple way to start your pitch or career summary is to write a short statement. This statement can be the opening paragraph for your career portfolio or resume. Complete Activity 5.5 to write your opening paragraph/career summary.

Setting Up and Sharing Your Career Portfolio

So, what goes into sharing a career portfolio? A career portfolio can be presented so that it appeals to an external audience, to potential employers. A career portfolio that you want to share (regardless of the platform used to present it) will generally be a website that you, as the author, could make public for others to view. It is up to you to ultimately decide how you want to share your career portfolio, but it could be a LinkedIn profile, your own website, or a website generated by the portfolio collection tool you have used during university. Your career portfolio will show the depth of your experience and evidence your skills and achievements.

If you are using a university virtual space to build your portfolio, remember to check how long you can use that space for. Many institutions enable you to keep using your portfolio space after you graduate. If not, it is

---

Activity 5.5: Write Your Opening Paragraph/Career Summary

Activity goal: Write a short summary of your career to use as the opening paragraph on your career portfolio, resume, or when talking to others.

Purpose and benefit: Preparing a career summary will help you to develop how you talk about your professional self and will be useful to introduce your career portfolio and/or resume.

Activity steps:

1. Write your opening paragraph that defines your career aspiration and summary/profile. Use the example shown in the section 'Pitching Yourself to Start Your Career Portfolio' to help you prepare your career summary statement. To do this, first describe who you are and your career aspirations, along with what you are studying.

2. Next describe two or three of your key skills. Finish the statement targeting the job or experience you are seeking.

3. Save your career summary as an artifact in your portfolio. And set a reminder in your calendar for 2–3 months to update your summary as you develop as a professional.

Takeaway: Your career summary will expand and develop as you progress in your career. By starting it now, you can practice your career summary while you are studying. Remember your career summary can be used whenever you want to present your career. This can include the career summary being the opening paragraph on the career portfolio you have prepared or as the opening paragraph on your resume or cover letter.

suggested you make a copy of your portfolio collection to use in another environment, such as making your own website. There are many free non-university hosting options available for those who want to continue to curate their portfolio once they have graduated.

Your portfolio website can be linked in a job application (resume/CV and cover letter) as a way of showing full detail regarding your career, experience, and achievements. Just remember that these links are not always accessed when used on applications, so you still need all the most relevant things in the application too (see Chapter 6: Resumes and Cover Letters and Chapter 7: Addressing Key Selection Criteria and Interviews on how best to produce these documents). You may also share a portfolio if you are engaging in freelancing activities, or with a company, to promote your abilities.

Importantly, what you share will not be the full collection of works and results that you may have collected over time. It's the portion that is useful to share with others at that point in your career. In the first couple of years of your career, this changes rapidly, so be sure to update between roles.

One portfolio approach is the LinkedIn profile. LinkedIn is well set-up to present your career profile, with an About You section where you describe your career summary. It also has sections summarising your work experience and skills, which is a great place to highlight formal skills and past work experience, and you can feature recommendations from others complimenting your skills or thanking you for work well done. Figure 5.6 is an example LinkedIn profile showing the about section.

There are additional pages where you can build sections that aren't always visible to others. Each section can highlight a different set of

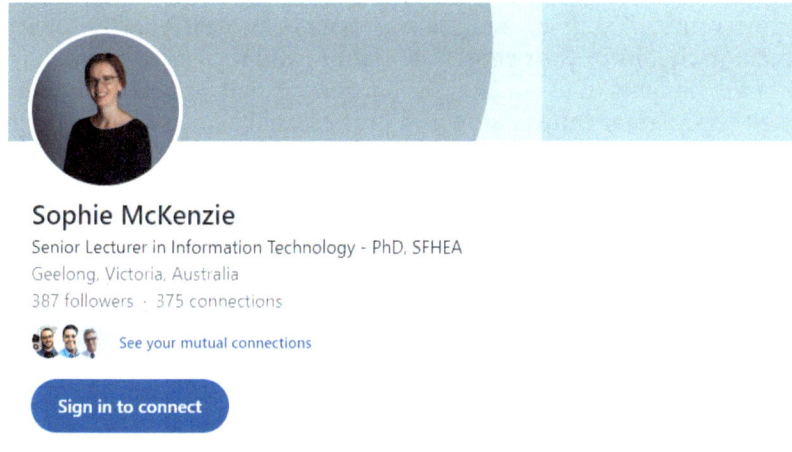

Sophie McKenzie

Senior Lecturer in Information Technology - PhD, SFHEA

Geelong, Victoria, Australia

387 followers · 375 connections

See your mutual connections

Sign in to connect

About

I am an experienced Information Technology Lecturer with a demonstrated history of working in the higher education industry. Skilled in Educational Technology, Training, Course Management and Curriculum Development. My research focuses on advancing computing education exploring how to support graduate employability. I also explore how immersive technologies support the human experience, particularly in educational contexts. While working as an academic, I also work as a career educator for Deakin's graduate employment division helping students across the university achieve career success and persistence.

My Deakin University profile provide more information and links to my research output: https://www.deakin.edu.au/about-deakin/people/sophie-mckenzie with my personal website demonstrating my projects further: https://tinyurl.com/yysu9zkb

Figure 5.6 An example of a LinkedIn Profile of the author.

skills or strengths, or highlight them in a different way. When you want to fine-tune the professional picture you are putting out to the world, you can just change out which panels are visible, and which aren't. While a website is perfect for some people, LinkedIn might be more accessible to others.

An Example of Displaying a Career Portfolio

To help you in preparing your career portfolio, consider setting up sections that will enable you to demonstrate, describe, and reflect upon your professional self and your achievements (as shown in the example in Figure 5.7). Having sections will highlight to employers your broader career goals, while also showing/evidencing your skills. Ultimately, it is up to you to decide what section to include and how much you'd like to share with others.

A career portfolio may take several different formats, but ultimately will describe yourself, your career goals, provide a professional summary (in relation to jobs in your chosen discipline) and showcase your skills. One of the benefits of using a career portfolio is that you can tailor and present your career information and professional self in a format that best appeals to you and your unique skills and abilities. In fact, making yourself 'stand out from the crowd' can be achieved through a portfolio, as you can tailor content presentation and style. To help guide you

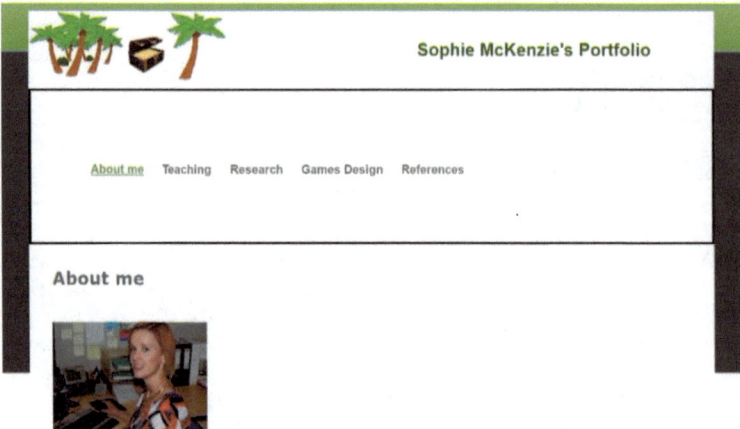

Figure 5.7   Example career portfolio.

in creating your first career portfolio, the following points outline some common elements that appear is most career portfolios:

1. **About you page:** This page can include a personal statement about you (your professional values, work attributes or type of work you like). A professional-style personal statement helps set the scene for the type of professional you are, without being specific to a particular job role. This section should also describe your education and employment history. Figure 5.7 shows an example of a career portfolio, highlighting the about me page.

2. **Career page:** This section should describe your career goals (or aspirations – see Activity 5.1). Use your career summary, as developed in Activity 5.5 as a way of introducing yourself and demonstrating the field you would like to work within. Your career page can help you demonstrate your awareness of jobs, and how generally you think your skills are suitable for job role(s). You can go into more detail on your skills in a later section (see the next point). Your career page will also briefly state your plans for career development in a professional manner. Having a plan will show that you know what you will do to achieve the goals as set in your career summary. It is important your career section reflects an awareness of the current the labour market. So, when including your plans for working after you graduate university, be realistic about how your career goals relate to jobs that are available. Activity 5.2 asked you to find out more about jobs that are available. Use this information to reflect on whether your career goals and aspirations are realistic in relation to what jobs are available in your area.

3. **Key skills page:** This is where you demonstrate your discipline and non-discipline skills, highlighting your skill set through artifacts and reflection. Using your portfolio artifacts, you have a variety of ways in which to present your skills, with opportunity to present groups of skills through a collection of artifacts and/or reflection as the evidence. In most environments, employers value interpersonal and communication skills as highly as discipline-specific skills when looking at a student's resume/CV.[5] But this can vary by environment, so be sure to do your research on an industry, discipline and location basis. Usually the more conservative the environment, the more likely they are to care about marks. Note that attitudes have changed in the last generation, so make sure you ask someone with knowledge about the workforce from the last five years. It is important for students in STEM to present a skills profile that is not just focused on discipline specific skills, but also considers transferrable skills. You may choose to focus on your major as a way of presenting your skills. Or you may focus on an experience (such as completing a placement) to highlight your key

skills. There is no one way to present your key skills, rather you focus on your unique abilities and use your portfolio artifacts and reflection to highlight these to others. Reflect on how you developed certain skills through your studies and other career experiences. Refer to Box 5.3 to see a reflection that could occur as part of a portfolio.

4. **Career achievements page**: What achievements would you like to showcase to highlight your career achievements to date? Did you receive a good mark on very relevant university project that really interested you? Have you been awarded or recognised for any volunteering work? Have you completed a placement experience, and what did you find out about your career by completing it? This section is to highlight any achievements that showcase your interest in your career. As you progress, especially after graduation, this portion will contain mostly employment achievements and only the most relevant achievements from your schooling (*e.g.*, an award in a relevant class, or a literature review on a closely related topic).

5. **An ongoing learning page**: While you are studying, it is expected that you will have a plan for future learning, such as what subjects are you yet to study and how these will contribute towards the development of your career. If you are exiting university, you will be expected to continue your learning, but this may be more self-directed than when you were studying. What continuous learning/professional development will you engage in during your studies, and once you have finished studying? How does this learning/professional development contribute to your career goal? You may even consider including your career experience plan in this section if you did not include details on this in a prior section of your career portfolio.

Your career portfolio should have an appealing design and consider user experience. How do those who look at your portfolio navigate successfully? Is there a consistent look and feel? Symmetry can help a lot with a design that is pleasant to look at. And unless you have design talent, keeping the colour palette simple is usually wise. Consider a design which maintains your details (name and email, keeping in mind who you might share your page with and privacy issues) throughout every page. You will need a design which also enables the user to navigate your content easily.

Now that you understand the multiple uses of a portfolio (collection, reflection and sharing), next you need to know how to leverage the things you have done during your time at university to present your portfolio. How do you use your assessments, completed as a course requirement, for your portfolio? What else do you put into your portfolio to enable you to prepare a career portfolio?

Confirming Your Professional Self

Now that you have the basics for developing your career portfolio, it is important you use and confirm the way in which you describe yourself. In other words, you need to confirm that the way in which you are describing your professional self makes sense with potential employers, leading into successful introductions, as shown in Figure 5.8. Two great ways to do this are through networking and careers coaching.

Using Your Portfolio for Networking

Networking is not just about making connections, but about making a meaningful impression and learning. It is important that, in addition to creating your career portfolio, you also consider how you promote yourself through other online professional sites, such as LinkedIn, to make these meaningful connections. Below we hear from Mark, who is an IT manager, about networking.

"One of the key drivers for anyone in the industry is networking. Networking does not just mean having connections on LinkedIn. Your networks that are of highest value are those where you have established a more personal connection", states IT Manager Mark. "It may be someone you have worked with in the past, or another contact who comments on the value and effort you put into your work. [...] It's important to consider your network regularly to see who you can connect with, and how to continue to build those relationships",

Figure 5.8　Networking and presenting your professional self; photo by Cytonn Photography on Unsplash.

states Mark. "As a hiring manager, I use my network a lot to understand applicants, looking at where they have been and what the feedback on them is. I use this information, along with how the candidate presents themselves and what I am sensing from the individual". Mark emphasises the importance of having a good network with personal connections to enable you to stand out. "These days, so much of the recruitment process is completed automatically through assistive technology. Unfortunately, these technologies will not be able to effectively filter applicants' unique qualities, the things only understood through human interaction. [...] And that's why your network is so important, as you're trying to reach someone to make an impression, to create a connection you can work with." A good network will facilitate "a person-to-person conversation to actually meet and greet and start the process", concludes Mark.

You can use your portfolio collection to help you showcase your abilities online, ensuring you establish meaningful connections to get a job. See Chapter 4 for details on how to develop your networking skills and a high-quality professional network. As discussed in the section 'Pitching Yourself to Start Your Career Portfolio', a summary or pitch will shape others' opinions and perceptions of you, so use your time at university to create your personal brand along with your career portfolio. Articulating yourself can be challenging, so it's important to start early and practice your career pitch (review the reflection section of Chapter 2 if you need help to articulate your skills). The information from your portfolio can be used to create your online personal brand, enabling you to present a consistent career narrative and engage in more successful networking opportunities. Below, Associate Professor La Fontaine talks about how critical networking is in her experience.

When building your career, Associate Professor La Fontaine states that networking is critical. "Past students have said to me that those words from me to network have stuck with them, and that's what they've done, and they just can't emphasise that enough when they talk to current students." Networking can take on many forms, whether it's "generating a LinkedIn profile or just not being shy to talk to people, ring up people, go to events, attend conferences", says La Fontaine. In addition to networking, she encourages students to leverage the skills and experience from whatever casual professional role they are engaging in, whether it's in retail or volunteer experience, as a stepping stone into the next professional opportunity. Networking and experience provide the opportunities that students can describe in their portfolio, but an important part of preparing a description of your professional self is "knowing how to articulate skills and experiences". La Fontaine advocates for "practicing the techniques to be able to articulate experiences, as they enable students to perform well in job applications and interviews".

Using Your Career Portfolio for Job Interviews

Your career portfolio may come in useful when you go to a job interview, particularly for a role you want to do post-graduation. You may have included your portfolio as a link on your job application/resume, making it something your potential employer is aware of. More often, you will have used it to practice how you articulate yourself in preparation for your interview, including relevant parts from your portfolio in your job application. Depending on the discipline you are applying for, there may be opportunities to demonstrate your skills and experience during the job interview beyond your verbal response to questions. This is a great opportunity to use your career portfolio to demonstrate your skills. Examples from a project, outcomes from your participation in a competition or hackathon, a demonstration of leadership skills, can all be emphasised in a job interview with the support of your career portfolio.

Validating Your Career Portfolio Through Career Coaching/ Counselling

Career coaching or career counselling is where you work with a specialist to explore your career options and career plan. Career coaching/ counselling can help you plan, clarify your career direction, and assist with marketing your skills and experience to potential employers. Career coaching/counselling provides you with self-management skills, which are vital to compete in today's job market. During your time at university, you can usually access a career coach through the university to help you focus on your career goals. Help comes in the form of group training and individual meetings. A career coach can also provide feedback on your career portfolio and on how you might best demonstrate your professional self. Through consultation, you can determine if your portfolio is doing a good job of demonstrating and furthering your career goals. You may need to revisit your career summary or re-think your portfolio presentation if your career goals are modified or changed through consultation. It is highly encouraged that you seek out career guidance at the university (if available) before you finish your studies, particularly as the service will most likely be free while you are a student! Also, check if the service is available after graduation.

Conclusion

This chapter has explored what it means to prepare a career portfolio. Along the way, techniques for exploring your career have been presented, enabling you to make the most of your portfolio collection. Creating a career portfolio is not a one-off activity, rather you will regularly revisit, revise, and refine your portfolio presentation, even after you have started in your desired professional role. It is important that building your career portfolio is considered an ongoing activity, as you continue to learn, grow, and develop as a professional. Even though your career portfolio might be 'in-progress' as you are still studying, your career portfolio can be used at any time to showcase your current skills, career goals, and plans as you develop your professional self. Just remember that a career portfolio is always a work in progress because each new year of work will bring new skills and experiences, achievements to highlight, and potentially new references or recommendations. Even the act of presenting yourself as a student-professional is great practice for presenting yourself to people in your future industry when you graduate. It is important you develop a career portfolio that suits your interests and showcases your unique skills for employment. The snapshots from those in the field throughout this chapter show you perspectives on how you can progress your career choice, and use your portfolio to evidence, reflect upon and articulate your professional self. Dr Coleman encourages all students to be curious and adventurous about their career, exploring multiple possibilities as they could be *"insightful and helpful, and allow you to do what you're passionate about, and for a job that might not even exist"*. Mark Giles adds that when hiring new graduates he *"looks at their approach to learning and their personality"* as *"personality plays as much into the selection of a candidate as their background and education"*. And *"your personality can lead you to a job that suits you best"*, he argues.

Curating a portfolio during your studies will enable you to think about your career in many ways. By collecting artifacts and reflecting on your skills and achievement, a career portfolio provides you with evidence that you can use in a job application. Having these artifacts available first enables you to reflect upon and better articulate both your personal and professional growth, ensuring you are best prepared for a happy and successful career.

Check Your Progress

In the first chapter you likely filled in a short survey to gauge your perception of your employability, which will have been emailed to you. This was called a self-perceived employability questionnaire (Activity 1.1). The subject on the email will have the title of the book in it. If you are ready to check your progress, use this QR code to recomplete the questionnaire and it will send the new results. Compare these results with any you have been sent previously by doing the following:

1. Bring the results up together and compare your results. If you are later in the book, you might have more than two. You can compare which results you want, or even all of them. Transferring the data to an Excel spreadsheet will help if you have many answers and are inclined to do that, or if you need an activity to help develop your Excel skills.
2. What are the differences? Sometimes you will feel a bit different on a different day, so not every value that changes by one point is a significant change. Maybe you had a good day or a bad day. Can you see any trends or big differences? Do any of the changes feel particularly true? Even if it is just one point.
3. Often when people learn more about a subject, they first loose a bit of confidence as they are starting to learn about how much there is to learn. Has that happened to you? This is an important process to be aware of. Maybe you aren't feeling as comfortable with your employability but that's because you had to realise how much there was to know to be able to go out and learn it. Well done, that was an important realisation.
4. Now that you have stood back and thought about your overall perceptions of your employability, are there any plans you think you might make or change? Or maybe something you have realised you want to prioritise or deprioritise? If so, make a note of these and set a calendar reminder once a week for the next month to ensure it doesn't get forgotten.
5. Enjoy the next chapter!

References

1. G. L. Ring, C. Waugaman and B. Brackett, The Value of Career ePortfolios on Job Applicant Performance: Using Data to Determine Effectiveness, *Int. J. ePortfolio*, 2017, **7**, 225–236.
2. S. Mckenzie, J. Coldwell-Neilson and S. Palmer, Career aspirations and skills expectations of undergraduate IT students: are they realistic? Conference presentation HERD, Sydney, Australia, June 2017.

3. K. Coleman, S. McKenzie and C. Wilkinson, *Developing Learning-Centered Approaches across the Discipline: Implementing Curated ePortfolios in Information Technology and International Studies*, The WAC Clearinghouse; University Press of Colorado, 2020.

4. D. A. Schon, *The Reflective Practitioner: How Professionals Think in Action*, Basic books, 1984, p. 5126.

5. Australian Government, *Graduate Outlook*, 2015.

Resumes and Cover Letters

ROSEMARIE I. HERBERT[a] AND
ANGELA ZIEBELL[b]

[a]Monash University, Faculty of Science, Australia;
[b]Deakin University, School of Life and Environmental
Science, Australia

In this chapter, we will delve into two of the most important written pieces that you will prepare as part of a standard job application. A resume (also known as a CV/*curriculum vitae*) and a cover letter work together to provide potential employers with a comprehensive overview of your skills.

The resume provides a detailed picture of your work history, education, skills, and accomplishments, while a cover letter more specifically highlights your key experiences and qualifications that make you a strong fit for the specific job you are applying to. Your cover letter should help to provide context for the information in your resume. Experiences that align clearly with a role but do not fit well into your resume (*e.g.*, editing a magazine or a small role in website building), can be highlighted in your cover letter. Your passion for the industry can be communicated most effectively in your cover letter. You do not know which of the two documents will be read first (or at all) and it is important to make sure that both (or three if they have asked for a key selection criteria response – see Chapter 7: Addressing Key Selection Criteria and Interviews) are adequately communicating your suitability for the role.

Overall, the key to making your resume and cover letter work together is to ensure they are both tailored to the job you are applying for. Tailoring your application documents is extremely important. Before we start constructing those two documents, we will talk about analysing a job advertisement so that you know how to identify what the advertiser is most interested in seeing in your application.

Analysing Job Advertisements

Although you might be tempted to apply to any and all jobs that you find (including some on a company's door as in Figure 6.1), you will quickly realise that applying for jobs takes longer than you expect. If it does

Figure 6.1   Have you seen any advertisements on doors lately? Think carefully about which jobs you will apply for. Photo by Eric Prouzet on Unsplash.

not take you longer than expected to apply (often hours per application), then you are not spending enough time to customise your application for each job. Put it this way, would you rather be ranked 25th in 25 applications or 5th in 5 applications? In the average job market, a shortlist for interviewing for most jobs will be around 5 (assuming that only one role is available). If you are 5th you are likely on the shortlist for an interview. If you are 25th literally no one knows your name. You probably passed the first test that determines whether you could do the job. This is an important step. However, without putting the time in to customise your application it is doubtful you will be offered an interview. We make this point in a very direct manner because it is so sad to see new graduates taking a volume approach in their first 6 months of applying for jobs and getting depressed and demoralised by rejections. We want to save you from that trouble by helping you work out the type of job that works for you (Chapter 8: Looking for a Job), helping you understand what the application needs and giving you some tools to present that application in a professional manner.

"I know so many other people are applying for like, 20–30 graduate programs and not even getting one... I just really made sure that it was really tailored. Every single word in the resume or the cover letter was actually relevant to that job that I'm applying for. So I think that really helps. That's all you can do."

– Heshan, science graduate working in the public service.

You will need to make sure that your resume contains the relevant keywords, and that the cover letter highlights the specific skills that are relevant to the company and its goals and values. This requires you to analyse each job advertisement and weigh up the suitability of the role – both in what you are looking for in a job and your fit for the job. Activity 6.1 will help you through working out this process.

Activity 6.1: Find and Analyse Job Advertisements

Activity goal: To understand how to analyse a job advertisement.

Purpose and benefit: You will have identified key transferable skills that you have in Chapter 2: Transferable Skills and Reflection. When you first start looking for jobs, you probably aren't sure of what roles interest you. Use these steps to identify jobs of interest and identify the skills that your industry is specifically looking for.

Activity steps:

1. Navigate to a job search website/platform such as Indeed or LinkedIn. Examine the parameters that you can choose. These typically include fields such as Keywords, Classification and Location from the landing page. You can also choose between different work types (full-time, part-time, casual) and a preferred salary range.

2. Pick your Location. Are you only interested in jobs locally, or is moving to another city (or country) an option?

3. Classification asks what industry or field you are interested in. You will want to select the relevant field(s) such as 'Science and Technology' or similar, *e.g.*, 'Government', 'Education and Training', 'Engineering', Healthcare', *etc.*

4. If you have a specialism that is listed (*e.g.*, chemistry, conservation, data science), you can choose a narrower classification. However, you will probably find it most useful to pick a series of related words/phrases that describe your education and industry, *e.g.*, biodiversity, ecology, biometric data, conservation. Pick two words to put into the Keywords field, along with the word 'graduate'. If you have trouble thinking of terms, visit a number of jobs and look for related terms in the advertisement. Look at the suggested jobs after your search and use some of those search terms and see what words appear in the jobs that are suggested for those too.

5. If you are still not getting many jobs of interest you can take off the higher-level selections that you put on in step 3. This can make the results messier because you will get lots of unwanted hits, but it may be helpful if you aren't sure where to start.

Continued

Activity 6.1: Find and Analyse Job Advertisements – (*Continued*)

6. Pick three jobs that sound interesting and write down the job titles. Open each in a new window or tab (or print them) and review the following sections.

- Job description or summary that includes the location, salary and the terms of the contract.
- Your responsibilities | About the role | Key accountabilities.
- About you | Selection criteria | To be successful.
- Who we are | About the company.
- How to apply.
- Position closing date.

7. Open your saved file with annotation software (or take your paper), and use the highlight or comment tool to pick out (1) What are the required qualifications or 'equivalent experience'? (2) What technical and transferable skills are required for the job? (Refer back to Chapter 2 if you have difficulty identifying these skills) and (3) What are the key duties and responsibilities?

8. This information tells you whether you meet, or almost meet, enough of the requirements for the role – and therefore whether you should apply to it. Write out which criterion you do not meet and which ones you are concerned you are borderline on. Is there anything you can do to fill that gap? Some gaps will be easy to fill (*e.g.*, brushing up on software usage or boosting your experience working in a group), while others will be hard to fill (*e.g.*, learning a language or having a different major). Some points will matter a lot (*e.g.*, if the company is bilingual) but other points employers can be more flexible around (*e.g.*, a mentor could supervise you focusing on a particular area you are weaker in or you could be assigned some training).

9. Skills are impossible to improve overnight so when you are looking at roles you should check if there is a pattern of where you need to improve. Did you spot any? Commit to a plan to improve a small skill or your experience over the next few months. What skill will you work to improve? How?

Takeaway: Remember that you do not have to meet all the criteria to apply (see the next section for more on this point). Keep the skills required for these roles in mind as you move through this chapter and put together your resume and cover letter. You should have looked at at least three different job advertisements and begun to work out what skills you need to succeed in the industry. You might have identified some areas where your skills are insufficient – flip back to Chapter 2 for some ideas on how to improve them.

Hint: There may also be a document with further information to download. This is often called a position description and holds more detail than the actual advertisement. It is important to look to see if this document is present, as sometimes looking only at the advertisement will put you at a disadvantage because all the details are not there. If there isn't, save each job listing as a PDF or similar. You can print the job listings but this is best avoided as you will be doing this a lot.

How Good a Fit Do I Need to Be Before I Apply?

This is a big question and it will depend a lot on what the job market is like for your environment. Remember that includes your geographic area, the industry you are looking at and the discipline you are applying in. Sometimes the employment market is just good in lots of places and a lot of industries. It is also important to keep in mind how your environment is at the time you are looking for jobs. You can also change your environment by looking for work in other industries or changing location. To gather information, you can use your networks, including (former) educators, family, former classmates and friends.

> *"When you're applying for roles you shouldn't hold back from anything, thinking that your [grade] isn't good enough... I was asked to supply transcripts when I was applying for roles at like really big companies like CSL [Commonwealth Serum Laboratories] or EY [Ernst & Young]... they just wanted to see that you'd attended."*
>
> – Ruth, biology graduate and manufacturing line manager.

As a definite rule, you will rarely be expected to fit all the criteria. Think of job descriptions as more like wish lists. The employer is putting something out there to recruit as many great people as possible and they will aim very, very high. Also, keep in mind that employers are adding to a team. They do not need you to be able to do everything on the list, but this is the sort of thing that the team does and they want as many additional skills as possible. It's fair to expect employers to train you in a few areas, but they won't consider someone who needs to learn everything on the job. As part of a team you can be useful in assisting more senior staff while you develop.

So where does that leave you? It is unlikely that you will get a role if you only fit half the criteria, but if you think it is the really important half, then it is worth a phone call to see what the person who put out the advertisement thinks. Reaching out to a hiring manager before applying could also help you to determine whether the job is a good fit for you, and/or avoid wasting their time and your time if your skills do not match what they were looking for after all. If you happen to know that the skills that you do have are especially in demand, an employer is more likely just to hire anyone and everyone with those skills. Generally, this will be accounted for in the advertisement, *i.e.* they will ask for less in an advertisement where some of the skills are hard to get. In these advertisements, you will see generous offers to help train people up or help get them certifications while still paying a high wage. This is rare, so if you happen to be in this category, congratulations!

On the other hand, never wait until you satisfy all the criteria before you apply. You are essentially overqualified at this point. First, this will mean that if you get the job, you already know how to do it all and you will not learn anything new. You are also likely to be bored in the role after a few months and even if you do not think you will be bored, the person reviewing the application might assume it. Usually when people apply for something they fully suit all the criteria for they are selling themselves short. Why are you not applying for roles where you would be trained in a skill or where practice would improve your abilities and experiences? Often, this is due to confidence. Try to touch base with friends, family, and career counsellors to check whether you are underselling yourself. If you have confidence issues generally, people who know you well will be a much better gauge of what level you should be looking for even if they cannot help you with the highly technical specifics around software, methods and equipment. You have worked hard to get your qualifications and now is the time to make sure you do not sell yourself short.

Confidence vs. Competence

Cultural norms can play a role in how confident you should be in your resume. Too much confidence can come across as arrogance, while too little confidence can make you seem like a less attractive candidate. Sometimes, to be confident is to be direct and assertive, while in other cultures, it is considered more confident to be humble and modest. You need to present yourself in a way that is most likely to be seen as confident in the culture of the company you are applying to. You will need to seek feedback from other professionals to gauge whether you are talking yourself up too much or not enough. Be conscious that this can vary based on gender and seniority in many places. Explicitly ask a range of people what you should do as a new graduate and why. If you ask a senior or very experienced professional, their suggestions may be a bit out of touch given your current level of experience.

You also need to be competent in the skills and knowledge that are required for the job. The best way to demonstrate your competence is to use keywords that are relevant to the job you are applying for. You can find these keywords by reading the job description carefully and by searching for similar jobs on online job boards. When you are writing your resume be sure to use these keywords throughout your document.

> *Hint:* If you are applying for a formal Graduate Program, your application is likely to pass through a range of automated checks before making it to a 'real person'. In this case, it's probably not a good use of your time to email or call to find out if the job is a good match for your interests and skills. Instead, you should focus on making sure your documents all read well and exactly match the language used in the advertisement.

Resumes

In the job market, your resume is often the first impression you make on a potential employer (although you can never be sure which document they read first). It's not just a list of your qualifications and work experience, but a reflection of your professional brand and a tool for showcasing your transferable skills. In this section, we will explore the key components of a successful resume, from formatting and content to tailoring your resume to specific job openings. We strongly recommend creating a brand-new resume with what you learn in this chapter, rather than trying to adapt one that you prepared previously.

There's quite a lot to cover in this section, so we advise that you address each subsection as a chunk and then take a breather. Specifically, the use of action verbs and creating bullet points of your most valuable skills will probably not come easily to you. Use each of the activities as a way to break up your construction of the document.

What Is the Point of a Resume?

Submitting a resume serves as an introduction to the employer and showcases your relevant skills, experience, and qualifications. It can help demonstrate why you are a good fit for the job and set you apart from other candidates. A well-written resume can increase the chances of being selected for an interview and ultimately getting the job.

Sometimes the only document requested by a hirer is a resume. If that's the case, it's your one chance to showcase your skills and prove you are the right candidate. It's really important to note that you'll need to optimise it for the specific job by using the same keywords as the job advertisement to make sure that it gets past the application tracking systems (ATS; see the next section). Most job advertisements require you to have a resume, but sometimes an employer will ask for a cover letter or response to key selection criteria. We will discuss writing key selection criteria responses in Chapter 7.

Applicant Tracking Systems

An Applicant Tracking System (ATS) is a software tool used by recruiters and hiring managers to manage the entire recruitment process, from posting job openings to screening and selecting candidates. When a job seeker applies for a job, their application is typically scanned and processed by an ATS to determine whether they meet the minimum requirements for the position.

An ATS uses a variety of criteria, including keywords, education, experience, and other qualifications, to evaluate each application and create a score or ranking. Resumes and applications that score well may then be reviewed by a human recruiter or hiring manager, while those that do not meet the minimum requirements may be rejected automatically or placed in a lower-priority pool for future consideration.

You should optimise your job application documents for an ATS by including relevant keywords and phrases from the job description and formatting your documents appropriately. It is critical to match the words in the job description **exactly**. For example, if the job advertisement is asking applicants to have 'good attention to detail', write that. While 'detail-orientated' means the same thing to human readers, an ATS may not pick this up. Also, avoid non-standard abbreviations and acronyms as these might not be recognised by an ATS. Remembering these points will increase your chance of being selected for further consideration by human recruiters.

What Is the Difference Between a CV and a Resume?

In many countries, the terms resume and CV (*curriculum vitae*) are used interchangeably (Table 6.1). They can also be used as slightly different versions of your work and training record. Typically, the main difference between the two is the length and amount of information covered.

A resume is a one to two-page document that summarises your work experience, education, and skills. It is intended to provide a brief overview of your qualifications for the employer to showcase your suitability for a particular job. It includes only the most pertinent information in terms of specific skills and experience.

Table 6.1 Naming and length of resume or CV conventions.

Country[a]	Document name	Document length
Australia (website 5)	Interchangeable	1–2 pages
Canada (website 4)	Interchangeable	1–2 pages
China (website 2)	CV (jiǎnlì – 简历)	1–2 pages
UK (website 3)	CV	2 pages
India (website 8)	Interchangeable	1–2 pages
Japan[b] (website 1)	CV	1–2 pages
Malaysia (website 9)	Interchangeable	1–2 pages
Philippines (website 7)	Resume	1 page
South Africa (website 6)	CV	1 page
USA	Interchangeable	1–2 pages

[a]For website details, see the list at the end of the chapter.
[b]In Japan, a CV is usually accompanied by a resume that is a personal profile (rirekisho).

A CV is a longer document that provides a more comprehensive look at your background, including a list of all your publications, presentations, and professional development activities. It is typically used in academia, research, government, and international job applications, and can be two or more pages in length. A CV provides a more detailed look at your entire career history.

It's important to note that the format and content of a resume can vary depending on the environment you are applying for. Fortunately, most jobs specify how long they expect your resume to be, and whether it is a resume or CV that is wanted. If it is not stated, give the recruiter or hiring manager a phone call or send an email if there is a contact. If there is no contact, you can just use whatever seems standard for that industry and geographic region. You might need to ask around a bit to find this out. Check a couple of internet sources such as Expatica or local career listing web pages to confirm what is required. You'll also need to consider what sections to include – we will discuss those in a bit more detail later in the chapter. This chapter will help you to construct a resume, rather than an extensive CV.

Formatting Guidelines

You have a lot of great experiences to share with potential employers, but the hiring manager needs to read your resume. Non-functional resume formatting can create a negative first impression for hiring managers. No one wants to read something that does not look nice and tidy. It does not have to be pretty or beautiful, but it does need to be practical and easy to look through.[1,2] Whatever formatting style you choose, it should not take the place of, or distract from, relevant examples.[3] It is critical that you share your document with others to help you identify any spelling mistakes, typos, or grammatical errors. Any errors make it more unlikely that you will be offered an interview because errors suggest that you are not conscientious or detail-oriented.[4,5] Once you have created your resume, let it sit for a couple of days and then come back to it. Does everything still make sense? Is it a good representation of your skills?

Before we jump into the layout of your resume, have a think about file naming conventions. You want to avoid naming your application documents as "resume". If a recruiter downloads all the applications for a particular job, they do not want to have to open every single file to find the one that you wrote. Instead, give your resume (and cover letter and key selection criteria responses) a sensible filename such as "Lastname-resume-jobnumber" or "Lastname-resume-jobtitle". The recruiter might need the job number, but you might want/need the title of the job to keep them all straight. This will also help you keep track of which documents you have used to apply to each job. The last thing you want is someone not being able to find your application when they want to check a detail. You might inadvertently drop off their list. In addition, it is also a nice touch that shows your professionalism, working to make the files easier for others to work with.

It is also important to make sure you submit the right file format. Check whether it is a PDF or file from word processing software that is required – if the recruiter cannot open the document, they are unlikely to chase you down, and therefore it is as if you didn't even put in an application. Only, it is much worse because you put in all that work and now cannot be considered.

> **Hint:** If you are being asked to submit a Word document, it can be a good idea to mark all the errors that the software is highlighting or underlining that aren't genuine errors by selecting 'ignore all' to avoid your reader seeing a bunch of red and green underlines throughout your otherwise amazing application.

Layout

Generally, resumes should not use text boxes to display or layout information. Using boxes can make it difficult for the reader to follow, there often is not an obvious box 1, box 2, box 3, *etc*. The reader might miss

SKILLS

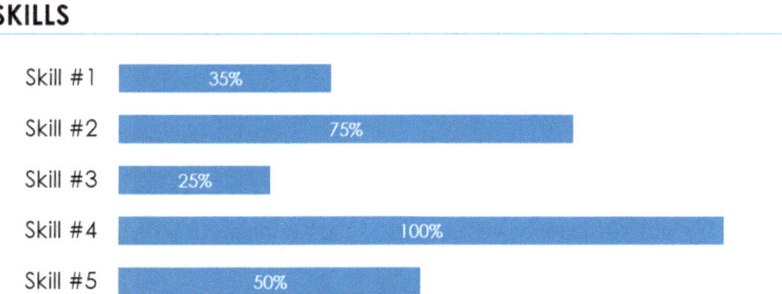

Figure 6.2 An example of skill bars. Avoid using skill bars or similar to indicate your competency in a skill. What does it mean to have a full bar of written communication and a three-quarters full bar of oral communication? Try to integrate your skills into the context in the body of the resume so that it is clear how, where and why you have developed skills.

parts of the resume because it is not in a linear manner. An exception may be using columns to display your skills side-by-side or listing your references. Tables are acceptable as they add to how easy it is to present fields and different aspects of your skills and experience neatly and in a logical, easy-to-follow manner. Also, avoid using 'skill bars' as described in Figure 6.2.

Remember that your resume is (presumably) being shaped to address science jobs. Keep the layout simple and uncrowded (you'll see a couple of poor examples later in this chapter). You aren't applying for artistic jobs where your resume acts as an instant portfolio of your artistic or design skills. There's no one right way to format a resume, but there are a couple of things that will affect it negatively, as discussed below.

Fonts and Colours

Your name should be at the top of the page, and stand out the most from the rest of the text, followed by each of the section headings. It can become visually overwhelming if you use too many different fonts. Use a maximum of two font types and stick to default fonts to ensure that they remain rendered correctly when your resume is opened. Use easy-to-read fonts such as Times New Roman, Arial or Calibri at no less than 11 points in size for section contents. If you are struggling to keep your resume down to 2 pages, you can use a sans-serif font at 10 points. Do not go any smaller than that though or your resume will lose readability.

For section headings, you could use a different font, or the same font in a larger size (14–16 points). You can also use **bold**, *italics* or CAPITALISATION to create impact. Do not go overboard with this.

Choose which you will do – you do not need all three. Be aware that screen readers sometimes have trouble with capitalisation so it can just be safer to not use capitalisation.

When choosing colours for a resume, it is important to consider the environment you are designing the resume for, as well as personal preferences. For some environments, a monotone resume is preferred. If you choose to use colours in your resume, ensure that these are consistent with your personal brand. Keep in mind that different colours may evoke particular emotions such as blue for trustworthiness or red for passion. You should stick to black for general text for readability. Shades of grey can be useful behind headings to help white headings stand out on a grey background, keeping the monotone theme but helping delineate the document. Here are two examples:

Career history

Career History

Personal Branding and Visual Appeal

Branding is a business strategy that uses colour and motifs to differentiate visually one company from another (see Chapter 3: Commercial Awareness for company branding). You can use personal branding to create a distinct impression on potential employers. Branding can be used to help recruiters link your job application documents together when they are choosing candidates to interview. Imagine that the hiring manager has twenty documents open on their screen – yours can benefit from standing out. You are aiming for yours to be visually attractive, but also not overwhelming or garish. It is important to note that this will vary with your environment. It is useful to talk to professionals working in areas you wish to apply to or to do your online research to fine-tune your approach after you learn the basics in this chapter.

You may also like to consider a 'theme' to your resume – if you are a conservation scientist you might use a rainforest as the header and a green shade matched to that image as your section header colour. You can use a basic header image to create a sense of cohesion with your LinkedIn profile, too. Overall stick to only one or two colours, otherwise the formatting may detract from the language that lays out your skills.

Margins and Spacing

Use a consistent margin size on all sides of the paper. We suggest a 1″ or ~2.5 cm margin – anything larger than this will make your information seem cramped on the page, and anything smaller will be visually distracting. You'll also want to make sure that your indents for each piece of information are consistent across the whole document to help keep the document neat and highlight that you are an organised person who knows how to prepare a good document. This is a skill all by itself.

Line spacing should be no less than 1.15 because less than that will make the information appear cramped and make it difficult to read due to a lack of white space. White space is the space between design elements that can help your content to stand out. It can also help to make text more readable and to focus attention on important elements (such as your name). By adding white space around text, you can make it easier for readers to scan and understand your content quickly. If you struggle to fill a 2-page resume, consider using a slightly larger spacing.

> **Hint:** If you are struggling to remain within a 2-page limit, think about what roles and responsibilities you have had and carefully select the ones that are most aligned with the job. Again, only include information that is suitable for the environment you are applying for. Less is not always a bad thing, particularly if the skills and experience you do have are a good fit for the role. You can likely condense the information a little more by:
>
> - merging your roles at one company
> - including only the most relevant responsibilities
> - putting text such as skill lists into tables
> - using bullet points to shorten sentences.
>
> Once you have a lot of experience you might like to create a 'master resume' or CV that has everything you've ever done, but you usually should not include all that information on a resume that you are submitting to a company.

Icons

You may like to include an icon for your LinkedIn profile, phone number or email address. There's no need to write next to each of these what they are because it is implicit in the information given (*e.g.*, myname@ mailservice.com is clearly an email address due to the @ sign). Alternatively, you can download your LinkedIn profile URL as a QR code that can easily be scanned. It can also be useful to ask other professionals in the area you are applying to as to what is seen as professional for that environment. Practise what you have learnt by doing Activity 6.2.

Activity 6.2: Critique Generic Resume Formats

Activity goal: To examine a range of resume formatting options to identify a style you like and could use for your own resume.

Purpose and benefit: You should start to understand what style of resume you want to put together and potentially identify a template to use to scaffold your resume.

Activity steps:

1. Open up writing software such as MS Word or Google Documents. From the Start menu, it will usually give you an option to search for a document type and select a template. Search for 'resume template' and glance over the options listed.

2. Open at least three of the resume templates and apply the following checklist:
   - What do you think of the colour choices? How many colours have been used?
   - Do any of the resumes appear crowded? If so, why?
   - Examine one of the resumes that use columns to separate information. Try reading from left to right while ignoring the columns. What kind of 'sentence' is formed?
   - Look away from your computer screen and then look back. What jumps out at you on the resume? *Hint: it should be the applicant's name.*

3. Type into a search engine 'example of a science resume'.

4. Select at least five and have two open at a time to compare them. Select the most appealing one, and then pairwise compare until you have a top and a bottom example.
   - Which person do you get a more favourable impression of?
   - If there are summary statements, compare them. Which is more engaging to read? Note down what words seem important/attention-grabbing.

5. Alter your search parameters. Search 'example of a bad science resume' and see what comes up. This should be fun. Use the above checklist and these additional prompts:
   - What have you discovered about the resume?
   - Why is it 'bad'? Is it the content or formatting that is the problem?
   - Is there anything redeemable about it? Do some of the bullet points have valuable skills? If they do, write them down as examples to refer to later.

6. Choose a resume template from step 2 that you found attractive and save it with the filename "Lastname-resume-draft". Edit it if appropriate to incorporate elements that appealed to you in step 4. Check that it meets the guidelines we have suggested earlier in this chapter.

Takeaway: You should have now had a look at about eight different resumes and begun to work out what yours should look like. You should have identified at least one style or template that appeals to you to slot your experiences in as we progress through this chapter. Remember that the style you choose should be appropriate for your environment. Always remember that the most important part of the resume is the content – you can always change the formatting later so just pick any that appeals at this point.

Two sample resumes are analysed in Boxes 6.1 and 6.2 to help highlight features of a resume. Once you have a draft of your resume you can also look through the notes in Boxes 6.1 and 6.2 to see if you have accidentally included any negative features.

Box 6.1

Continued

Box 6.1 – (Continued)

1   The applicant's name should be clear. In this example, the S/S is distracting from the applicant's name (Sam Smithson).

2   In most environments, you shouldn't include a physical address. Usually, a city/county gives sufficient location information.

3   There is no particular need to include your Skype username unless asked for in the application. It may be useful to include a website address if you have a portfolio of relevant documents or codes.

4   Skill bars are generally not helpful. What does it mean to have a 6/10 for Japanese, but a 9/10 for communication?

5   The majority of environments do not expect you to include your secondary/high school education as it's irrelevant once you become a graduate.

6   Your education should include the university/college that you have attended, any majors/minors that you have and the location. It's also important to include an expected graduation date so that employers know when you will be finished with your studies.

7   A summary or objective statement should include specific details about your goals and suitability for the role.

8   The layout of the document is lopsided, with unequal distributions of white space. The columns also make it difficult to work out where to look next.

9   This experience correctly includes the relevant details of job title, company name, location and dates worked. However, this experience is in a paragraph format, which is inconsistent with the other experiences.

10 This experience has a mixture of paragraphs and bullet points. It's also missing details of where the company is located. Additionally, it is listed as 'current' which does not indicate how long the person has worked in the role. Any current roles are usually listed first under the experience section (reverse chronological order).

11 This experience uses action verbs to describe the tasks associated with the role. However, the font style and size are different to the other experiences, which adds to the overall unpolished resume presentation.

12 While hobbies can be useful to include in a resume if they add depth or additional skills, they should be professional. 'Sleeping' is not an appropriate hobby to list.

Box 6.2

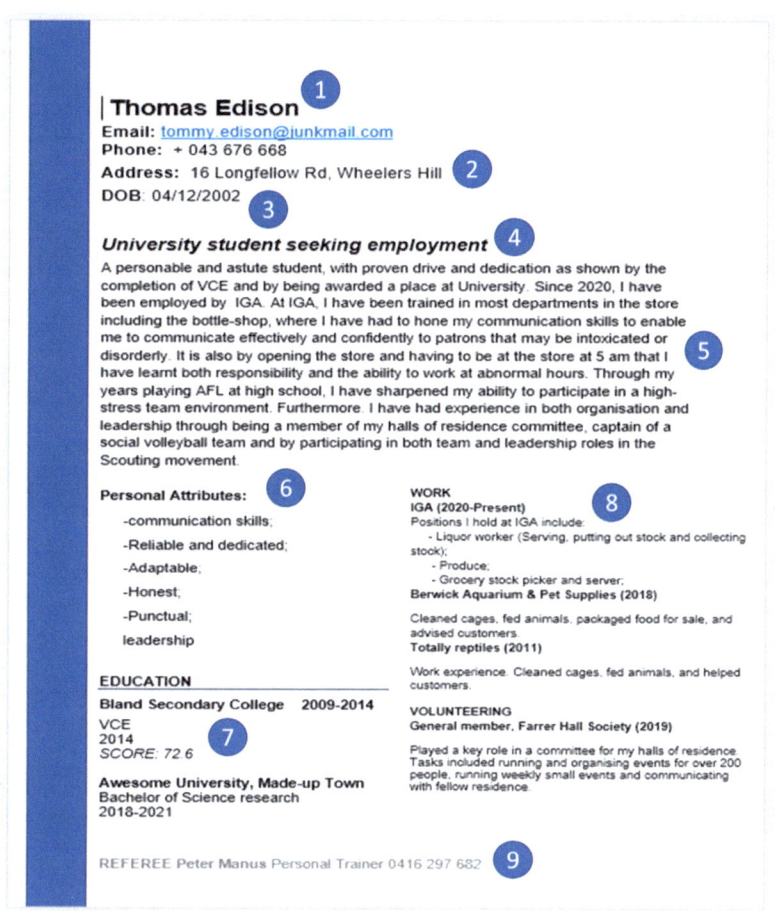

| Thomas Edison ①
Email: tommy.edison@junkmail.com
Phone: + 043 676 668
Address: 16 Longfellow Rd, Wheelers Hill ②
DOB: 04/12/2002 ③

University student seeking employment ④

A personable and astute student, with proven drive and dedication as shown by the completion of VCE and by being awarded a place at University. Since 2020, I have been employed by IGA. At IGA, I have been trained in most departments in the store including the bottle-shop, where I have had to hone my communication skills to enable me to communicate effectively and confidently to patrons that may be intoxicated or disorderly. It is also by opening the store and having to be at the store at 5 am that I ⑤ have learnt both responsibility and the ability to work at abnormal hours. Through my years playing AFL at high school, I have sharpened my ability to participate in a high-stress team environment. Furthermore, I have had experience in both organisation and leadership through being a member of my halls of residence committee, captain of a social volleyball team and by participating in both team and leadership roles in the Scouting movement.

Personal Attributes: ⑥

-communication skills;

-Reliable and dedicated;

-Adaptable;

-Honest;

-Punctual;

leadership

WORK
IGA (2020-Present) ⑧
Positions I hold at IGA include:
 - Liquor worker (Serving, putting out stock and collecting stock);
 - Produce;
 - Grocery stock picker and server;
Berwick Aquarium & Pet Supplies (2018)

Cleaned cages, fed animals, packaged food for sale, and advised customers.
Totally reptiles (2011)

Work experience. Cleaned cages, fed animals, and helped customers.

EDUCATION

Bland Secondary College   2009-2014
VCE
2014 ⑦
SCORE: 72.6

Awesome University, Made-up Town
Bachelor of Science research
2018-2021

VOLUNTEERING
General member, Farrer Hall Society (2019)

Played a key role in a committee for my halls of residence. Tasks included running and organising events for over 200 people, running weekly small events and communicating with fellow residence.

REFEREE Peter Manus Personal Trainer 0416 297 682 ⑨

1   The applicant's name should be eye-catching. The font for Thomas Edison is smaller than that of (4) 'University student seeking employment'.

2   In most environments, you shouldn't include a physical address. Usually, a suburb/county gives sufficient location information.

Continued

Box 6.2 – (Continued)

3   Generally, you should not include your date of birth as it can lead to ageist bias and also be an information security risk.

4   The statement included here doesn't give any further context to the application and also adds to a picture where the applicant is still a student.

5   When you read this section, you'll see that there is a lot of language that is unprofessional. A summary statement should be concise and personalised to the job you are applying for. It is optional to include this section, and it may be more useful to use the space in your resume to more thoroughly talk about the skills and experiences you have.

6   These bullet points are not consistently formatted and some of the skills listed are not appropriate for a professional environment (e.g., honesty).

7   This applicant has included their high school/secondary college, and also not provided sufficient detail about their university studies.

8   The formatting and content of the experience section are inconsistent. A mixture of different fonts, formatting styles and bullet points *vs.* sentences detract from the visual appeal of the document. The heading for 'Volunteering' does not stand out and is lost.

9   In most environments, you should not include a personal referee. Referees should be from professional contexts and it should be clear what their relationship is to you.

Action Verbs, Dynamic Verbs or Strong Verbs – Illustrating Your Skills

Maybe you have an existing resume and you're beginning to think that this chapter is irrelevant to you. But have you heard of action verbs before? Action verbs are a type of verb that express what you do and convey a sense of ownership of whatever is occurring. This section will introduce you to action verbs to take your written communication skills to the next level. At the end of this section, you will feel confident to implement your action verb knowledge to improve your job application documents and LinkedIn profile.

What Are Action Verbs?

Verbs are a valuable part of writing that are inherent to communication in English and usually go somewhere in the middle of the sentence (or at the end). Verbs are sometimes referred to as 'doing words' because they describe the activity being undertaken. Action verbs in particular are used to create a sense of ownership of an action. These make you seem as if you engaged with the activity, rather than just passively letting something happen to you. For example, which of these statements seems more compelling to put in your resume?

1. At my old job I had to make new reports with some old data to show the best sales technique for the product.
2. I *analysed* pre-existing data to *create* interactive reports that *showcased* the best sales techniques for our top three products.

Hopefully, you answered that the second sounds more engaging, and you noticed that in it the writer appeared more "powerful" or more in control of what happened. Not only have action verbs been used in this example (analysed, create, showcased), but a bit more background information has been included to provide context for the actions. Yet, it is not a longer explanation because using strong action verbs is space and length-efficient. Writing with action verbs gives useful context to your skills so that the potential employer doesn't need to imagine how you will fit into their business.

Examples of Action Verbs

Great action verbs to use include: Created, Designed, Developed, Enabled, Facilitated, Delegated, Managed, Operated, Identified, Advised, Delivered, Coached, Consulted, Analysed. 'Utilise' is not an action verb and doesn't add meaningful information to your statements. 'Use' works perfectly well, but you should not be using it in your resume.

Let's take a look at another example before you try writing your own in Activity 6.3. The applicant has included a list of activities that they perform at their current job:

- I covered front-of-house service.
- I made sure the till balanced correctly.
- I was responsible for talking to other people to share tasks.
- I had to work in a team.

See how these do not clearly explain what is unique or important about these activities? Here is an example where action verbs (delivered, ensured, negotiated) and fragment statements illustrate a clear example of the abilities of the candidate. We've also included a couple of positive words (exceptional, accurate, fairly) to further illustrate the successful completion of the tasks.

- Delivered exceptional front-of-house service.
- Ensured accurate money-handling practices.
- Negotiated with colleagues to fairly distribute tasks.

Activity 6.3: Use Action Verbs

Activity goal: To begin identifying and using action verbs while writing about your prior work responsibilities and experiences.

Purpose and benefit: Upskill your ability to use action verbs and build a foundation for your resume to better capture what you did in your role.

Activity steps:

1. Think about a responsibility or role that you have in your current position and write at least three sentences that describe it. Or, look at the list of bullet points that you wrote down from Activity 6.2.

2. Look at the verb in each sentence and consider the following points:
   - Where is the verb located? It should usually be at the beginning of the bullet point.
   - Analyse the verbs you have used. Are they action verbs?
   - Have you used the same verb more than once?

3. Rewrite each sentence as a bullet point with the action verb rules in mind. At this point, it doesn't really matter what action verbs you use (you can choose some from the previous page).

4. Find a job advertisement that has several skills-based criteria that you meet. What are the key skills you've learnt in each role that are relevant (for broad categories see Chapter 2)? Adapt the bullet points that you wrote in step 3 or write additional points that cover the skills.

5. Think about how you have developed skills in each of these areas. Then, try to use action verbs that mirror the language used in the advertisement. As we discussed in this chapter, the keywords must be the same as in the advertisement as well. If it says 'excellent verbal communication', you need to write that, not 'confidently delivered information orally'.

6. Give yourself some space and put away your hard work for a day or two. Then come back and read your work afresh. Is what you have written actually what you do in your role? Do the keywords match the advertisement? Have you reused the same action verbs? A quick internet search will help you find some more action verbs to use as well.

7. Once you have written up a couple of roles, ask someone else in your environment to look over your action verbs and bullet points. You can also ask other professionals to share their resumes with you to give you a feel for the environment as well.

Takeaway: You are hopefully feeling quite comfortable with using action verbs to improve your writing. You should have a couple of points about roles you have had in the past that match a specific job advertisement. Remember that you'll need to adjust them for every job you apply for.

Resumes are not the only place to include action verbs. You should also use these in your cover letter, key selection criteria responses and LinkedIn profile. Let's look at a cover letter example first. A common mistake made by applicants is to use the following phrases. If you see "I was able to use X", "I managed to do X", or "I believe I show X", then that's an ideal place to use an action verb. An additional advantage is that you reduce the word count – giving you more room to include other accomplishments or to increase the use of white space.

Always remember that people reading your documents need to be 'snagged' early – but once you have them engaged, they want to hear about the details. Make sure to personalise your cover letter, resume and key selection criteria response for every job you apply to.

> *"If you show that you've actually looked at the company... that they can see that you've gone a little bit further than just copying and pasting the different company name... [you'll be] much more successful."*
>
> – Ruth, biology graduate and manufacturing line manager.

Despite the advice we've just given you, do not get too excited and over-use action verbs. Putting your work aside, or asking someone else for their opinion will help you determine this. Remember that your writing needs to mirror the keywords in the job advertisement as well. The action verbs are there to persuade the reader, but they shouldn't overtake the actual content of your resume.

The Components of a Resume

Below we detail the main considerations for your resume. However, it is really important to remember that this is baseline information. Different environments (different disciplines and geographies) have a variety of approaches. Some of these are very small differences, such as a preference for a slightly longer or shorter resume. Other points are specific to a particular set of countries (*e.g.*, including a photo or a lot of personal information) or disciplines (*e.g.*, portfolios in engineering or IT to demonstrate your products). Some are obvious (*e.g.*, asking for your resume in both languages in a dual-language company) while others are not (*e.g.*, some individuals and companies just love the idea of a one-page resume). Some will depend on how well you know the environment (*e.g.*, exactly how formal you should be and what specific formal language words you should use to show respect).

So, you must take the advice we give you here and put it into context before you use it. If you are feeling worried or overwhelmed, work through Activity 6.4 to help you. We will teach you how to write directly

Activity 6.4: Identify Your Worries About Making a Resume

Activity goal: To become comfortable with the idea of a professional resume and familiar with some of the resumes in your environment.

Purpose and benefit: A resume should show off your best skills and needs to fit the environment you are trying to be hired in.

Activity steps:

1. Find a geographically local person you trust to ask about the requirements for a resume. They might even share theirs with you. Be conscious that the person you ask should be at a similar career level to you. If you examine your manager's resume for example, you might find that the language used is too 'braggy' or confident for a new graduate.

2. Consider your answers to the following questions.
   - What worries do you have about creating a resume? Why?
   - Are you concerned that you are bragging about your accomplishments? Why?
   - Do you think that your current skills are not relevant? Why? What might you need?
   - Are you afraid to apply to jobs because you think that you aren't experienced enough? What informed this feeling?

3. Speak to your selected person about your own resume once you have constructed it. That doesn't mean that they need to do a line-by-line read – it might be as simple as asking whether you should include a photo or whether to include your specific address.

Takeaway: You should be ready to more confidently start writing your resume using the instructions in this chapter. You'll also potentially have a person lined up to proofread your resume for you once you've created it.

---

to be concise and firm. If your environment does not suit that style of language, you can use the same learning to tune your directness the correct amount to get the type of writing you need. It's important to know how to ask about the level of directness in your writing, and how much you should proudly "claim" your success in your application. So, we write here about how to write firmly about your achievements because we know this is something that science students and graduates often struggle with. It is up to you to explore how much you need to employ that skill when you apply to a given environment.

This probably sounds like too much at once, but do not worry, you won't have to address all these considerations. Most people only have to think about a few. A student returning to China from North America will need to understand that the North American university probably doesn't give career advice specific to China. A junior engineer needs to understand that engineering companies can have their own culture and when moving into consulting there might be new considerations that come into play in the world of consulting. Understanding of these environmental differences can be developed through conversations and online research, plus information from career pages and your university. The following lays out some basic tools. We recommend working through the chapter to gain general awareness and then coming back to using specific sections as you develop each piece of your resume.

Contact Information

Your name and contact details must be clear and easy to spot on your resume. This can be done in a couple of ways, but as a minimum, your name should be at the top of the first page and in a larger font size than the rest of the text. You can make it a contrasting, dark colour to make it stand out. That being said, do not use fluorescent yellow or green as this is more distracting than helpful. Your name should also be in the header of subsequent pages.

Next, include your phone number and email address, as well as the link to your LinkedIn profile. Usually, you should not include your physical address to the level of your apartment or house number; however, if your last job or educational institution is located in a different place from where the job is, then you can include your location or state a willingness to relocate in your cover letter. Note that in some countries, putting your exact address is normal. If your exact details are not needed, you should leave them off to protect your privacy and make sure that you are protecting yourself from identity theft. The nature of the type of information shared during job applications (*e.g.*, address, birthday, photo) makes resumes a common target for people mining personal data.

Personal Information

Other information that is not usually needed in many countries is your gender, date of birth, nationality (and visa status/working rights), marital status, religion, and a declaration of previous income. Including such elements is highly dependent on the environment you are applying to (but we have summarised a couple in Table 6.2 for you). Look at a couple of the large online job listing company websites (such as Seek, CareerOne, *etc.*) to determine what is typical for jobs in your target country. You can also consult a trusted friend or colleague for what is considered standard in each context. International students from a different background than yours can also be a valuable source of global information.

Table 6.2 Contact and background information that may be included in a resume. Remember that these are generalisations, always check.

Country[a]	Photo	Street address	DOB, visa, government-issued number, gender	Civil/ Marital status	Religion	Current income/ Expected salary
Japan (website 1)	✓		✓	✓		
China (website 2)	✓	✓	✓	✓		✓
UK (website 3)		✓				
USA/ Canada/ Australia (websites 4 and 5)						
South Africa (website 6)		✓	✓	✓		
India (website 8)	✓	✓	✓			✓
Philippines (website 7)	✓	✓	✓	✓	✓	
Malaysia (website 9)			✓			

[a]For website details, see the list at the end of the chapter.

In most environments, you should not include a photo. However, in some countries, you should include a professional headshot on your resume as well. One drawback of including a photo is that it may inadvertently cause biases. Employers should base their hiring decisions on qualifications and skills, rather than physical appearance. Additionally, some employers may have policies against requesting or considering photos in the hiring process to avoid any form of discrimination.

Hint: Make sure your email address is appropriate – smartalec.is. cute.2000@aol.com is not going to give an employer the right impression of your professional self. Poorly chosen email addresses can jeopardise your chances of being shortlisted for an interview.[6,7] Try to create an email address that is just your first and last name, *e.g.,* firstname.lastname@gmail.com. Avoid including your birth year in the address as this could cause an ageist bias or, worst case, be used in identity theft.

Objective or Summary Statement

If needed in the environment you are applying to, the Objective Statement of a resume contains a brief overview of your career goals and the type of position you are seeking. It should go just under your name and contact details. It should be no more than 2–3 sentences long, and it needs to be specific to the role you are applying to. You should aim to use the same language that is used in the job description – if you are applying for a customer service-based role, weave that into your objectives. If you have a lot of relevant experience for the role you are applying for, you may wish to leave this section out to give yourself additional room to elaborate on your previous positions. If all you are including is how you want to work in the job you are applying for, the statement is a waste of space. It can be a good idea to include this section if there is no request for a cover letter. While the 'About' section on LinkedIn is a little different to a summary statement in a resume, it can be a great place to get a feel for what to write if you are stuck. Here are two examples to look at, and then use Activity 6.5 to write your own summary statement.

Here's a bad example (although it could be worse):

> "I am a science student with great communication skills and a passion for research. I'm looking forward to joining companies who also have a passion for science and are forward-thinking in the area of data science."

A better example:

> "Highly motivated and results-driven young science professional majoring in Pharmacology and Data Science. Three years of experience in customer service at a family-owned pharmacy and an internship at [**Consulting Company**]. Seeking a graduate position to further develop my skills and contribute to data-driven decisions."

Education

Education should always be listed in reverse chronological order. You should include your highest degree and the name and location of the institution. This is often done well with a table if you have more than one educational entry to list. If you are currently studying, you should include your expected majors, and month and year of graduation. This is particularly important if you are applying for next year's graduate programs that ask you to have graduated within a certain number of years. If there is an area in which you have completed relevant coursework to the position you are applying for, you can include this here. For example, if you are applying for a role at an IVF clinic, you may want to include that you studied developmental biology. Depending on the environment

Activity 6.5: Create a Summary Statement for Your Resume

Activity goal: To examine other people's summaries and "About" sections on LinkedIn to help you write your own. If you are not planning on using a summary, then it is still a good idea to try this activity to see if you can learn more about writing and how other people communicate in job application documents.

Purpose and benefit: Create an engaging generic summary statement for your resume.

Activity steps:

1. Open LinkedIn. Look through your connections (first or second), and find at least two professional profiles (not student ones). If you don't have many connections yet, go to one of the groups you have joined and look at some profiles from there. Ideally, they have careers that are similar to the ones you are considering, but any professional will give you an idea of what to write.

2. Write out how each individual includes their skills. Are these incorporated into sentences, or left as a list? What structures do you like or dislike? Why? Are there any that would work for you with additional fine-tuning? Do not just copy and paste them into your document.

3. Are action verbs used? (Details earlier in the chapter.) What action verbs? When you see this person use them, does it feel like they are bragging or just confidently stating their experience and achievements?

4. Put your degree details, top experiences and strengths into an AI text generator and ask it to produce a summary statement for a resume. You will find that what it writes is extremely broad, but if you are struggling to start writing a statement about yourself it can be a good starting point.

5. Try taking what the text generator gives you and adjusting the text by asking the text generator to change the tone of the writing. For example, is it too bold or too meek? You can also ask for the text generator to use formal language (e.g., "I would appreciate it if you…" or "Let me know at your earliest convenience.") or less formal to make it longer or shorter, etc.

6. Reflect on what you liked about the different summaries you read and write down the three key messages that you learnt that you want to try to use next time you need to generate a summary for your resume.

Takeaway: You should have around three to four sentences about yourself to include in a resume if necessary. Remember, though, to tune these depending on the job. If what you have created is not company-specific, it could be used for your LinkedIn "About" section.

you are applying to, you shouldn't include your grade average or every subject you have taken – just the subjects that are relevant to the role you are applying for. If an employer wants a complete breakdown of your subjects and marks, they will ask for your transcripts. This is not that unusual with newly graduated applicants, but over time it is getting less common in most places. In a lot of Western countries, the request for transcripts is just required as official proof that you got the mark that they use as an eligibility cut-off for their applications.

> **Hint:** Do not include your high school or earlier unless this is typical for the environment you are applying in. Depending on the high school (think a private, expensive school *vs.* a school with a bad reputation) you could be projecting an outdated persona that doesn't come across as professional. You could be discriminated against for going to too nice a school or not a nice enough school. Just do not give people the chance to do that if you do not need to. If you have significant leadership or volunteering experience at your high school, consider including these under a 'Volunteering' section.

Work Experience (and Volunteering)

A list of your past job titles, the name of the company, dates of employment, and key responsibilities and accomplishments. Depending on your history, you may have a lot of past jobs, but some fit best in proving that you can do the role you are applying for. Roles could include:

* customer services such as food service or retail (Figure 6.3).
* internship either as part of your degree or independently.

This is the section of your resume that you should spend the most time working on. Each job needs to have a list of key responsibilities and achievements as bullet points, not paragraphs. These should usually begin with action verbs. Try to use a range of words for each job you have done, but do not go too heavy with synonyms. Make sure that it is still readable. You probably want a maximum of three to four bullet points per experience, but this may vary based on how many experiences you have and how long you worked in the role. Try to choose a diverse range of experiences and tasks to show off your versatility and skill set while also remembering that having a similar skill or experience appear more than once is OK. It reinforces that you have had a lot of experience in that space and you can highlight different aspects of similar experiences to fully describe your skills and experiences or responsibilities. Remember

Figure 6.3   Experiences in the service sector can be included to demonstrate examples of working in teams, managing difficult customers, and potentially commercial awareness. Photo by Alex Block on Unsplash.

when you are applying for a specific job, make sure that the language exactly matches what is used in the advertisement. If the job ad says they want you to have 'excellent teamwork ability', you state that you have an 'excellent teamwork ability' as well as detailing what it specifically was *e.g.,* coordinating a small team of six people or leading a group of 20 people.

You should aim to include three to four bullet points per experience, and these should be in order of importance/relevance to the job you are applying for. Try to quantify your accomplishments with numbers (*e.g.,* worked in a team of 10, served up to 300 customers a shift) to help with specificity. As you gain experience, you will need to weave roles together. Maybe you have had two similar internships and you can present them together. Or, you have had four different roles at the same company. Do not just keep adding examples. Only include the more relevant skills and experiences – ideally the ones that are very similar to the role you are applying for. If you have internship experience that closely matches what the job is looking for, it makes sense to give this experience the most detail, and less on your work at a retail store. If there is a sales or customer service element, make sure you clearly outline the skills you developed working student jobs, such as waiting tables, serving in a café, or working a customer helpline.

Before you revisit Activity 6.3 to refresh your memory on how to create bullet-pointed descriptions of your skills, here's an example of a paragraph about a role.

I worked with students from grade 3 to 12, as well as TAFE students, in Biology, Chemistry, English and Math using an online tutoring system and whiteboard to communicate. I worked with fellow tutors in order to present a cohesive educational front and expand one another's knowledge. I also needed to maintain student and fellow tutor confidentiality at all times.

Here's that same role written as bullet points with action verbs (note 60 words for the paragraph and 40 words for the bullet points).

- Tutored primary through tertiary students in Biology, Chemistry, English and Math
- Facilitated online interaction through utilisation of a virtual whiteboard
- Collaborated with a team of 15 tutors to provide real-time solutions
- Maintained student and tutor confidentiality

If you have a slim resume or gaps in your employment history (*e.g.,* you were studying, or took time off for family commitments), **volunteer work** can help to fill these gaps and show that you have been active and productive. The volunteering section of your resume shows that you are interested in something other than money. Including volunteer work on your resume can show that you have a well-rounded, balanced life and that you have interests outside of work. Volunteering demonstrates your commitment to a cause or organisation and shows that you are dedicated to making a positive impact in your community. This can be particularly appealing to employers who are looking for employees who share their values.

Alternatively, you may have volunteering experiences that could be interpreted in a biased manner. You can be more generic in your description, *e.g.,* coordinate local community charity fundraising efforts, and include it in your cover letter rather than your resume. If you do not have any volunteer experience, remember that it doesn't need to be a sustained effort – a once-off one-day commitment still helps you to develop your transferable skills. Over time, as you gain more industry-specific experience, you may need to remove your volunteering experience because your resume is becoming too long. If this does happen, you can still talk about those experiences in your cover letter or the interview, even though they may no longer fit in your resume.

Here are some ideas of where you might have developed skills that don't necessarily fit within a specific work role. We'll help you think of other examples to match transferable skills in Activity 6.6.

- Tutored other students or family members? Maybe you've helped family members adjust to going to university or moving to a foreign country. Possible linked skills include initiative, intercultural competency and the all-important communication.

Activity 6.6: Identifying What Skills You Have and Where You Have Developed Them

Activity goal: To reflect on your skills and experiences to date to identify relevant skills to be included in your resume.

Purpose and benefit: To identify what transferable skills you have developed through your experiences and education.

Activity steps:

1. Keep in mind each of the key transferable skills that were discussed in Chapter 2.

2. Consider the following prompts to help you identify experiences you have that do not specifically fit into a paid role. You will likely have to think hard to come up with an extensive list. Often things that you take for granted are not done by many people. For example, you might have looked after or taken a close family member to appointments. Maybe you house- or pet-sat for someone. These show your humanity and time-management skills. Not everything is going to be suitable for a resume, but you'll have a nice selection of things to choose from.

3. Make sure that you use action verbs and note at least three examples where you have developed or used these skills. You might have noticed that you have lots of skills and examples that do not necessarily fit well in a resume format – do not worry. You'll be able to use these in your cover letter and interview.

4. Once you have identified your skills you might have recognised that you have a lot of skills in one particular area, such as teamwork. Or, maybe there's one that you are lacking. Some skills may be more suitable for specific job niches. You will need to seek out other activities to build your skillset if it doesn't match what current job advertisements are looking for.

5. Think of the activity as giving you a 'master list' of your skills. Remember that this list will continually update as you gain more experience – you'll want to come back and revisit it from time to time.

Takeaway: You now have a list of examples with linked skills to help you build your resume. It's quite empowering to think about the things you have already achieved – you'll probably be surprised by the number of things you have done.

- Walked dogs, or volunteered at an animal rescue? You've likely developed teamwork and time management skills and shown initiative.
- Play a team sport such as soccer or rowing? There's the obvious skill of teamwork, but you also probably had to communicate effectively and coordinate some parts of the commitment.

Certifications and Licences

A list of relevant technical, professional, and soft skills that demonstrate your abilities and strengths. This section can be valuable if you have particular coding languages or computer software skills that do not fit under your education, work or volunteer experience.

You should always include your language proficiency in any languages you speak. This section is of particular importance if you are applying for roles in non-English-speaking countries. Additionally, including languages you speak is a convenient way of showing that you have intercultural competency. You should include the level of your proficiency – 'beginner proficiency', 'working proficiency', 'native proficiency', *etc.* If you have grown up learning the basics of a range of languages, this field might be overlooked, as you are surrounded by others who are more skilled. However, just because your aunt is not impressed by your skill in her mother tongue, do not let that fool you into excluding your proficiency. Being a weaker native speaker means you still have a great working proficiency that will help with greeting someone in that language, making a client feel welcome and culturally safe, or travelling in a region that speaks that language.

Include any relevant certifications or licences you hold, such as a first-aid certificate (Figure 6.4). If you've completed LinkedIn Learning modules or tests, this is the place to include that information on your resume. Include relevant standardised language tests scores, such as IELTS (International English Language Test) or HSK (Chinese Proficiency Test) if the languages are relevant to the country you are applying for.

Awards, Scholarships and Achievements

This can include any awards you have received for your work. Avoid the temptation to include awards and achievements from when you were in high school unless they were important and related to something you are still doing. If you won an athletics scholarship at the end of school and you are listing skills from coaching roles you have had since then, listing a big scholarship might help round out the picture. You should only include awards if they are particularly relevant to the job you are applying for, such as 'Sales Associate of the Month' if applying for a sales

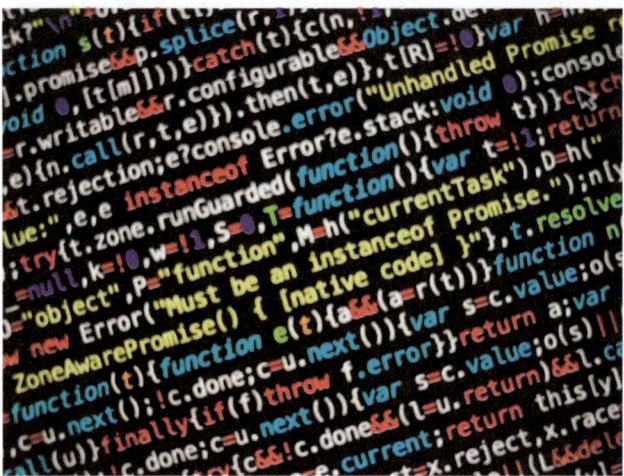

Figure 6.4 Some ICT and other skills can be captured as certificates outside your degree. Photo by Max Chen on Unsplash.

role. Your resume should present your best self as a graduate and old information can make it seem as if you have not achieved anything more recently. You can exclude this section from your resume if you have nothing you want to brag about.

Memberships

If you are an active member of a professional organisation or affiliation, you can include this. It's particularly valuable if you have completed an accredited university major with an associated Society, such as Chemistry (Royal Society of Chemistry), Physics (API) or Psychology (APA). If you are not a member or if you do not know anything about memberships, it is a good idea to see what there is for your discipline in your area. Professionals and people who graduated a few years before you can help you understand this space for your environment. There can be fees for being a member, so at times you might choose not to join. Memberships are often cheap or free while you are a student. If you are still a student, you can investigate any potential memberships now, look for mentors and connections to future workplaces, and learn more about different career paths and options in your environment.

Hobbies

If you're a student with very little professional experience, you may be struggling to make your resume two pages. Depending on the environment, you might find it valuable to include your hobbies to demonstrate transferable skills that are valuable in the workplace, such as leadership, teamwork (think of any organised team sport), communication, and creativity. These can also give a sense of your personality and what you are like outside of work. The hobbies you include should still be professional-feeling and constructive. Things such as 'hanging out with friends' or 'watching TV shows' aren't appropriate as they do not add any insight into who you are in terms of your character and you as a potential future employee.

If your hobbies and interests are related to the industry or field you are applying for, they can show your passion for that area and demonstrate your commitment to it. This might include your scuba diving if you are applying for a marine biology position, for example. Remember, though, that your resume shouldn't be too long, and you should prioritise the most important information, such as your education and work experience. Some examples you might include are:

- **Sports:** Playing sports or participating in fitness activities can demonstrate physical ability, teamwork skills, and a healthy lifestyle.
- **Creative pursuits:** Painting, photography, or music can demonstrate your creativity and artistic skills.
- **Reading and learning:** Including a list of books you have read or courses you have taken can demonstrate your intellectual curiosity and your desire for lifelong learning.

Figure 6.5 Hobbies can help you demonstrate leadership, strategy, dedication and more. Photo by Hassan Pasha on Unsplash.

Hint: In some countries, like the USA, extracurricular activities, such as volunteering and hobbies, are essential to include in a resume and these experiences build a clear picture of the applicant (Figure 6.5). In contrast, including hobbies and interests on a Chinese resume is less common than in Western countries. If they are included, it's best to list hobbies that give evidence of skills – such as chess (problem-solving) or team sports (teamwork). If you add hobbies, they should add to the cohesive story of who you are.

Referees (and Reference Letters)

A referee typically refers to a person who can provide a recommendation for the job applicant. These people may be former or current supervisors, colleagues, professors, mentors, or other professionals who have worked with you closely. This person can vouch for the applicant's skills, experience, and work ethic, and can provide insights into their character and suitability for the position. The referee can write a reference, often on an online portal. More commonly, a referee will be called or emailed by the recruiting company. Importantly, the referee should be someone who communicates well (both written and oral as it might be a phone call, email or letter).

The referee does not always have to be senior to you and you should ensure there is a good match between the job you are applying for and what the referee is likely to be able to speak to. However, in countries where hierarchy is more important, ensuring that you have higher-ranked referees can be important. You should always check what is the best approach in your environment by reading, talking to career educators where you can, and talking to friends, family, former classmates, and colleagues.

Always ask someone for permission before listing them as your referee. We'll walk you through the process in Activity 6.7. It can be very disconcerting to receive an email or phone call unexpectedly, and there is a good chance that your referee won't speak to your best qualities optimally if they are not prepared. Most of the time, three referees should be enough. In some cases, employers may request that applicants provide a certain number of referees or specify the types of referees they would like to see listed on the resume, such as former supervisors or colleagues. It is important to follow these instructions carefully and choose relevant referees. If it's been a while since you interacted with them, offer a coffee date or similar to catch up (this is a great opportunity for you to do a little networking). This will give you a chance to talk to them and refresh their memory on your best abilities. You can also check whether their current contact details are correct.

Activity 6.7: Select Referees for Your Job Application

Activity goal: To identify at least three people whom you have a professional relationship with who would be happy to provide you with a referee.

Purpose and benefit: Identifying the best people to include as your referees.

Activity steps:

1. Go through your resume and identify each role where you have worked in a team or reported to someone. List those people in order of those you are most comfortable with asking to act as your referee. You also might have a mentor (see Chapter 2) you can ask for advice. You want each reference to be able to represent a different aspect of your abilities.

2. Write out the following five pieces of information about your referee: name, company name, job title (which hopefully shows their relationship to you), email and phone number.

3. Check whether there is a good spread of skills. If something is covered more than twice and is not central to the role, ask yourself if you might be better off swapping out one of the referees for someone else who can talk about a variety of skills.

4. Remove anyone that is a family member or a personal friend.

5. For those of you who do not have enough references, jot down the names of a few other people who you are less confident about asking to be your referee.

6. Talk through the list and the suitability of any candidates with a few people in your environment. What ideas did you get? Do they make suitable referees? What did people think? Is this a confidence issue, or a lack of referees, or both?

7. If the main problem is reconnecting with someone and not asking them out of the blue, set time aside in your diary to write them an email or call them and catch up. A casual chat with a former colleague or supervisor is the perfect place to ask if they will be your referee.

8. Put together your final five people for your referees list. What have you learned from this activity? Do you have enough good referees? Do you need to plan some relationship-building to turn past colleagues into possible references by developing a better relationship? If so, review Chapter 4: Networking and picture yourself networking more to extend the circle of great referees you have (remember that there is a networking format for everyone).

Application: Once you have identified referees, you might have recognised that you have a lot of great people to evidence your skills. Or, maybe all your referees are coming from the same position. Hopefully, you now have a clearer view of your situation and a way forward to develop a great set of referees.

Takeaway: You now have a list of referees to include on your resume. Remember that you should check in with them every six months or so to make sure that they still remember you, and are happy to act as your referee.

It can also be common not to include your referees in your application and instead include a line that says, "please contact for referees". When you do this in an environment where it is the norm, you will get contacted when you are shortlisted for an interview, or even after the interview, when the company is reviewing applicants. By not including the referees, you protect your referees' privacy and you find out whenever you are a candidate of genuine interest. This helps you time a check-in with your referee, potentially even after the interview when you can give your referee more information about what it is that they need to speak to. Your referee will never lie, but in keeping to the most relevant points, they can ensure that your potential future employer knows as much as possible about you to make an informed decision.

In line with this, if you've had a successful job interview and the panel tells you that they want to contact your referees, this is a good sign. You should contact your referees (that you have previously asked to be listed – Figure 6.6), and inform them of the context of the role, and the company so that they can attest to your most relevant skills. If you are applying for a customer service role, you do not want your referee to talk about how you work well independently. Also, provide your referees with any documents that are relevant (such as your resume) or may be helpful when they speak to the hiring manager.

A general rule is that for local referees you should list a phone number and for overseas referees you should just include their email address, but including both for each is fine. The rationale behind this is that an

Figure 6.6 Make sure you notify your referees beforehand and make sure they are going to speak positively for you. Photo by Dustin Belt on Unsplash.

overseas referee is probably not going to be available to give a reference in the middle of their night. Alternatively, you can just ask your referee how they would prefer to be contacted or give both or leave it up to the hiring manager to decide what is appropriate. Note that hiring managers often resist contacting overseas referees in preference to your local referees as the process is more convenient and the referees in-country have more familiarity with local processes and needs. If this might cause you issues, for instance, your best referees are overseas, let the hiring manager know that. The hiring manager can consider it when contacting them.

For most roles, three referees are a good number to include. A useful approach is to have around five different referees that you can select from depending on the job you are applying for. Carefully consider the relevance of each person – this will again depend on the environment you are applying to.

- Are you expected to show off your diversity of roles/skills by listing referees from a wide range of different jobs?
- Is this an opportunity to name-drop that you have a contact within the company you are applying to or someone who is a key name in that industry?
- Should your referee be a peer or colleague, or someone in higher management?

If you do not have five people you feel comfortable asking, you could consider getting more involved in extracurricular activities or volunteering as a good way to build those personal connections. If you feel embarrassed to ask, just remember that everyone will have had to ask a referee to vouch for them at some point. Be as confident as you can – if someone says no, it's not necessarily that they are unwilling to support you, it may just be that they do not have any spare time at the moment. This is particularly true if you are asking your referee to provide you with a reference letter, which takes more time, thought and effort than a 15 minute phone call. Or they feel like they do not know you enough or in a way that would make them a good referee.

Reference Letters

A *reference letter* is a document written by a person (a referee) who knows you professionally and can attest to your skills, abilities, and character. The purpose of a reference letter is to provide additional information about you beyond what is included in your resume or application and to help the recipient make an informed decision about your suitability for the position or program. A reference letter typically includes details about your work ethic, communication skills, strengths, accomplishments, and other qualities that make you a good candidate. It can also provide insight into your character, such as your reliability, integrity, and professionalism. These days it is also common for the comments to be captured in a referee's report when the referee fills out a form, which can also have sections where they rank the applicant's skills. The report is just a different way to capture similar data.

A final note on resumes is to make sure yours is always up-to-date. It's often easiest to add your more recent job when you have just been hired, as you can take some experiences from the job advertisement to help you detail each role.

> *"Always keep your resume up to date, because you never know when you're going to be giving it out to people."*
>
> – Ojasvi, chemistry and physiology graduate and medical research assistant.

Cover Letters

A cover letter is a one-page document that is sent with your resume to provide additional information about your qualifications and interest in a specific job. Its purpose is to introduce yourself to the employer and explain why you are a good fit for the position. Keep in mind that you can never be sure of what document order hiring managers are going to read now that applications are dealt with online. Some hiring managers may skip over the cover letter – but that doesn't mean you shouldn't have one. Other hirers may read only the cover letter.

A cover letter can also showcase your written communication skills. It is important to keep your cover letter concise, professional, and free of errors. Errors include grammatical, spelling, typos and formatting inconsistencies. Use the cover letter to provide additional insights and details about your qualifications. Do not use bullet points because the cover letter is your chance to show off your written communication skills. If English is not your first language, get at least two other people to examine your work for spelling and grammar issues. It may be tempting to use generative AI to write your cover letter – in Activity 6.8 we have asked you to try it out. Be concise and precise. Link each position to the relevant part of the job advertisement. The hiring manager may only spend 60 seconds looking over your cover letter – what is the bigger picture you want them to see?

Activity 6.8: Trial AI to Write a Cover Letter

Activity goal: To create an AI-generated cover letter and critique it.

Purpose and benefit: You will recognise some of the limitations of AI in preparing job application documents. This will also help you identify what doesn't work well and why, to help you avoid making similar errors in your own cover letter.

Activity steps:

1. Use an artificial intelligence text generator (e.g., ChatGPT, Google Bard) to write a cover letter. Give it an instruction such as 'Write a cover letter for applying to [the name of a job you are applying for] at [X company]'. Ensure that you give the AI platform instructions like the length, the skills you want to highlight and the tone that you want.

2. Critique what is generated by assessing the following points:
   - Is the letter addressed correctly to the company contact person?
   - What skills are mentioned in the cover letter? Do you have these skills? Are these skills mentioned in the job advertisement?
   - If the cover letter has described some past jobs that 'you' have done, are these true/accurate? What kind of language is used to describe 'your' actions?
   - Does the cover letter give a convincing reason why you should be hired for the role?
   - What type of language is used? Is it formal, informal or does it use odd-sounding words? Does it sound like you wrote it?

3. You will probably find that the cover letter(s) you have generated are extremely vague, and also very USA-centric. The AI program is unlikely to use action verbs to create a concise summary and probably includes a list of skills with no contextual information. These skills also may or may not have been listed in the original advertisement (because you can't feed it the advertisement). Additionally, even if the AI has listed some details about the company, there is no guarantee that this information is current because AI text generators often only have access to older data.

4. Refine the cover letter. Give the AI program instructions to include at least five action verbs. What type of words does it suggest?

5. Check the cover letter against the original advertisement. What has been done well, or what can you adapt to help write your cover letter?

Takeaway: AI is capable of writing cover letters, but it is not particularly good at it because it lacks creativity. Cover letters are an opportunity to showcase not just your qualifications and experience, but also your personality and communication skills. AI doesn't know anything about you personally unless you give it those pieces of information to work with. Resist the temptation to use it for writing your cover letter – it's not specific enough to give a sense of your personality and create a good impression with a hiring manager.

Hint: Use the job posting or job description to guide your writing. Make sure you address all of the qualifications and requirements listed in the posting.

- What are my qualifications?
- What are my goals?
- What intangibles are important for me to fit in at a company? *e.g.,* values.

If you don't have all the skills required for the job, be willing to learn new skills. Employers are often willing to train new employees. We will talk about how to address this in Chapter 7 when responding to key selection criteria.

Letter Header

You should include your details, including your name, phone number, email address and LinkedIn profile. This information should either be centred in the header (with your name a little larger to grab the reader's attention) or at the top right of the page. If you are using graphic branding, make sure it matches your other documents. To get an idea of what a typical cover letter looks like, work your way through Activity 6.9.

Somewhere at the top of the page, you should include the date you are applying. You should include the company name and company address to show that you have done your homework. If the company has no physical address, you can include the company's URL homepage instead. Then, include the job title and/or reference number. Address the letter to the recruiter or hiring manager's name. If there isn't a contact person listed, you can ring the contact number on the advertisement to find out. Do not be afraid to call them – showing initiative can be your foot in the door to an interview.

Paragraph #1 – Introduction

Most of the time, you should lead your cover letter by stating the job you are applying for and how you learned about it. This is where you can share any connections you might have to the company, such as a current or former employee of the company. Or, you might have someone to 'name drop' that the hiring manager knows (if you happen to have this connection). This small paragraph is where you can put forward your eagerness for the role and very briefly give an example or two of how your key experiences match what they are looking for.

Hint: You may be thinking, where can I find information about the company? You can find relevant information about the company in the job advertisement itself, the company website or annual report, or a LinkedIn page. An internet search in any browser about the company, any key people and their products will also improve your knowledge. You'll want to be aware of competitors, market trends, demographics and areas of current innovation. You can follow the business on LinkedIn, or even reach out to the recruiter directly through the platform. Revisit Chapter 3 for more information on how to find information about a company and its industry.

Paragraph #2 – What Can You Do for the Company?

The second paragraph of your cover letter should highlight the skills and experience you have that directly align with what the company is looking for. It's important to tailor your cover letter to the job you are applying for by showing how you meet the requirements of the position. Use the same language as is used in the key selection criteria section of the job advertisement.

If you use a broad statement such as "With my experience and skills, I am an excellent candidate for the position", you'll need to follow it up with evidence. Stating that you believe something is not the same as giving evidence that you know how to do the job. Avoid giving a long list of skills that you have that match the job description. Instead, pick the top three to four skills from the job advertisement and give concise but detailed examples of when you have used these skills. If possible, summarise several different experiences you have, and perhaps include things that didn't fit in your resume but are relevant to the job. Employers want to see or hear specific examples of how you have used your skills. Use quantifiable data to demonstrate the impact you have made in your previous roles. You might include your skills, knowledge and experience, but also your attributes (*e.g.*, patience, flexibility) and motivations. Remember that the hiring manager may only look at one or two of your documents before deciding to invite you for an interview – your cover letter needs to provide enough information that it can stand alone from your resume to some extent.

Activity 6.9: Critique Generic Cover Letters

Activity goal: To examine different cover letters to identify strengths and weaknesses of different styles, both in writing and layout.

Purpose and benefit: You should start to understand what to include in your own cover letter and the language that is appropriate for a cover letter.

Activity steps:

1. Type into a search engine 'example of a STEM cover letter' and look for those that come from your environment – that's your geographic region and your discipline plus industry. Select at least four cover letters and have two open at a time to compare them. See if they follow the structure we have suggested.

- **Introduction:** Where was the role advertised, what is it, and what is the applicants' connection to it? Does the applicant appear keen to apply for the role and not too over-excited or, alternatively, unengaged or uninterested?
- What can the applicant do for the company? What unique skills or experiences make them a good fit for the role?
- How does this applicant align with what the company wants? Are there shared values or needs?
- **Conclusion:** Has the applicant included their contact information, enthusiasm for the job and alluded to future interactions?

2. Remember that the cover letters you have examined for this activity are unlikely to be paired with a job advertisement and you therefore won't be able to compare the language between the two. Nevertheless, consider the following questions:

- What aspects of the cover letter do you like, and why? What do you think would be useful to include in your own cover letter?
- Do any of them include specific details about the company that is being applied to? If so, what has been focused on?
- What elements are engaging to read? Remember that although this is a chance to show off your writing skills, it should still be concise.

3. Try 'example of a bad cover letter' to see some poor examples and give you an idea of what to avoid. Use the above checklist and these additional prompts.

- Why is it 'bad'?
- Is there anything redeemable about it?
- How would you check that you haven't made any of these mistakes?

Takeaway: You should have now had a look at several cover letters and gotten a good idea of the type of language you are expected to use in one. Keep in mind what you liked or didn't like and why as you write your own.

Hint: Most application tracking systems (ATS) check for matching words in the cover letter and resume to the key selection criteria and/or job description. If you are updating a cover letter from a past job application to the company, make sure that the language *exactly* matches the current role. We've emphasised this to you in multiple sections of this chapter because it is critically important to get it right. Remember that something such as 'hard worker' may be referred to as 'has a good work ethic' – you need to write exactly what is in the current advertisement you are addressing.

Avoid statements such as "I believe I am the right person for your company to hire". You need to do more than just 'believe' you are a good fit – you need to prove you have those skills and explain where you developed them.

Paragraph #3 – How Do You Align With What They Want?

This section of the cover letter needs to frame what specifically you have with what the company wants. Avoid needy language such as "I want to work for you because I believe that you will give me an opportunity to build my skills." Instead, try to frame your current skills as something you have that is going to benefit the employer and show the employer that you are headed in the same direction as they are. You will want to keep in mind any benefits, perks, or opportunities for growth and advancement that the job offers. Some will be obvious and are easy to comment on in a positive and forward-looking manner, *e.g.,* you want to train up to use a particular programming language or type of instrument and the role involves being trained up to higher levels as you work. Other benefits you may not want to mention but might be very important like good insurance, holidays or pensions. You can potentially frame this as you are attracted to the company, as it is clear from its benefits scheme that it looks after its employees and as a loyal and reliable person you like to see that your loyalty will be returned.

An example of this would perhaps be wanting to live in a particular area where there is a gap in skilled professionals in a specific field. Perhaps you want to live where you grew up and the testing company that you are applying for is based in the small country town of your childhood. If the job requires you to relocate, you can talk about how you are keen to experience other environments/broaden your experiences.

You should explain why you are interested in the job. Is it your dream job, or have you always wanted to work at this company? Why are you applying? The research about the company you have done will support these points – do your values align with the company? If the job advertisement or similar has a "Apply if you are excited to..." section, mention those things. There is a lot to address in a cover letter, so thinking about the exact points you want to make and why is important, as is concise wording. This will take a lot of practice and redrafting, so give yourself time, especially if this is your first time applying for work. This is a great time to have an appointment with a mentor and/or career advisor from your institute to help you draft and redraft your cover letter.

You learned about action verbs and how to use them in a resume earlier in this chapter. Now you need to practise using them in a cover letter to add impact to your statements. You need to use full sentences to show off your writing skills.

Which of these two statements seems more compelling?

1. "I recently graduated with a degree in Science. I'm really excited to work for your company because you are at the forefront of your field."
2. "Dulux has a reputation for innovation and premium quality products. As a recent Bachelor of Science graduate majoring in chemistry, I have developed skills in HPLC that would enable me to thrive as part of your company. I am excited about the opportunity to contribute to a company that prioritises social and environmental responsibility because I am passionate about sustainability."

Hopefully, you selected the second one. See how the first is quite vague and could apply to almost any job that an applicant is interested in? The second showcases the candidate's commercial awareness and their research into the company. Practise your use of action verbs in a cover letter in Activity 6.10.

Paragraph #4 – Conclusion
End the letter by expressing your enthusiasm for the position and reiterating your interest. Provide your contact information again (*i.e.* list a phone number or email address) and allude to future interactions. This can include an offer to provide additional information, and/or answer any questions the employer may have at an interview.

Signature
Close the letter with a professional closing such as "Sincerely" or "Best regards" and sign the letter by hand (Figure 6.7). If you are submitting

Activity 6.10: Practise Your Use of Action Verbs in Cover Letters

Activity goal: To convert vague skill statements to specific examples with action verbs to create impact.

Purpose and benefit: Develop your ability to use action verbs and build a foundation for your cover letter.

Activity steps:

1. Write two to three generic skill statements or use the below example to start with. For example, you might want to describe your ability to communicate in writing.

2. Identify where you developed this skill. Perhaps you have written a literature review or a marketing report.

3. Break down the skill into smaller steps and use action verbs. You could use the terms 'research topics', 'outline my ideas', and 'create visual summaries'.

4. If possible, quantify your results to show the impact of your skills. Provide specific examples of how you have used your skills to achieve results.

5. You can follow this sort of formatting for your own cover letter paragraphs. You should include the role where you developed these skills, and match them to what the current job advertiser is looking for. Here are a couple of examples for you to rewrite if you get stuck.
   - "During my studies, I gained valuable experience in conducting research and experiments, analysing data, and interpreting results. I also developed strong skills in communication, teamwork, and problem-solving, which I believe would be valuable."
   - "I have always been passionate about this industry, and I am impressed by the way this company stands out. I am excited about the prospect of contributing to the development and improvement of your products, while also supporting the company's values."

Takeaway: Cover letters need to use strong action verbs and include specific details of the company you are applying to. You should now have a couple of examples to work with in terms of the skills discussed, but you'll need to do some research into each company that you are applying to in order to make your cover letter powerful.

Figure 6.7   Always end on a strong and positive note; photo by Signature Pro on Unsplash.

the letter electronically, you can type your name or use a digital signature. Follow this with your name (only if signed) and contact information (usually a phone number). Yes, you've included this information at the top of the cover letter, but you want to make it as easy as possible for the hiring manager to contact you.

> *Hint:* It's always useful to have a digital signature on file. You can then use this for signing PDF documents and other items. You can even use it to sign your contract after you successfully land a job. You can either put your signature onto a piece of paper and photograph it into a digital jpeg, or create a signature image using a stylus on a device (trying to use a mouse or trackpad to create a signature is quite difficult).

The level of formality in your cover letter should be appropriate for the environment you are applying to. This is also true of your resume and key selection criteria – the tone that you use to write these documents needs to be professional, but appropriate. For example, if you are applying for a job in a big consulting firm, you should use a more formal tone that mimics that of their job advertisement. However, if you are applying for a job at a start-up, they might use more casual wording in the advertisement, which implies that they are casual and friendly and would welcome a casual (but not unprofessional) application. It is possible to be too respectful in some cases. In particular, you need to check that the way you have introduced or signed off is respectful. You will find this hard to judge for yourself, so it is important to ask others for their advice.

Activity 6.11: Critically Examine Your Cover Letter

Activity goal: To look at your cover letter in the context of a job application.

Purpose and benefit: You should have an outsider's view of your cover letter.

Activity steps:

Now that you have written your first cover letter, take a step back. Let it sit for a bit if you can (1–2 days), and then come back to it. Be as objective as possible. Pretend you are a recruiter and address the following questions:

- What is the main message you get from the cover letter in 60 seconds?
- Does it interest you? Does it make you want to speak to the person that sent it?
- How does the applicant come across? Does it seem like they are bragging too much about their skills? Or is it not clear what they have to offer?

Takeaway: Remember hiring managers and recruiters are people. While it's important to make sure they get concrete examples of your skills, they also do want to get a sense of you, and whether you will fit into the company culture.

What If No Cover Letter Is Required?

If no cover letter has been requested by the potential employer, it is good practice to still include one if there is a place to do so in the application portal. Many people hate writing cover letters because the letter must be specific to the employer (and it's necessary to be specific about skills).[3] Including a cover letter will make it easier for the recruiter to get a sense of your personality, which is likely not evident in your resume and Key Selection Criteria responses. You'll stand out in a good way by including one.

If there is no place for you to upload a cover letter, consider including a career objective or summary statement in your resume to give better context for your examples and also showcase your writing skills.

What If the Application Is via Email? Should I Attach a Cover Letter?

If you are applying for a job *via* email (rather than directly through Seek, LinkedIn, or other job boards), your first email can act a little like a cover letter. Depending on the instructions in the job advertisement, you may want to shorten your 'cover letter' to the bare basics that will make you stand out from other candidates – think of it as a teaser for the rest of your documents.

LinkedIn Is Your Not-so-secret Application Weapon

Do you have social media profiles? What happens when you use a search engine to look up your name on the internet? Employers often look at social media profiles when short-listing applications.[8,9] You can use this to your advantage by having a professional, active and well-maintained LinkedIn profile (Figure 6.8). Conversely, a terrible photo (think rude or disrespectful) or a messy unprofessional profile online can give a bad first impression and possibly prevent you from being hired.[10]

If you have lots of jobs, certifications or skills that you do not have room for in your two-page resume, LinkedIn can act as a 'reservoir'. You can keep all of your experiences and simply show or hide those that are relevant to the current jobs you are applying for. With this in mind, you should make sure that the content included on your resume is consistent with that on your LinkedIn profile. The LinkedIn profile can verify what you are saying in your application document –

Figure 6.8   LinkedIn is a powerful professional tool when used well, even for students; photo by Souvik Banerjee on Unsplash.

particularly if someone else has endorsed your skills. Your mutual connections also prove that you worked for that company, and who you worked with.

LinkedIn can present the most positive version of yourself if you spend the time to get it looking professional. Additionally, LinkedIn is a valuable tool for job hunting and provides access to a vast network of professional connections, employers, and job postings. Some positions on LinkedIn have an 'Easy Apply' option that uses your profile information to prefill job applications – make sure everything is up to date before applying in this manner. You'll need to double-check that the language used in the LinkedIn advertisement matches what has been pulled from your profile information, otherwise it's not worth your time to click apply.

Some helpful tools of LinkedIn are:

- **Professional networking:** When at a networking night, you can whip out your LinkedIn profile to connect with your new contact. You'll then be able to investigate their career history and connections, and even ask someone to introduce you to another person of interest. You can also use LinkedIn in the reverse – to find out about someone's history before meeting them in person. See Chapter 4 to refresh your networking know-how.
- **Company research:** LinkedIn provides a wealth of information about companies, including their size, industry, and recent news. This information can be used to personalise your cover letter if you apply to one of these companies.
- **Job postings:** LinkedIn has a large database of job postings and can even be used to apply directly to employers. LinkedIn includes job postings from other countries by default, so you can have a look at what is happening overseas from your current location quite easily.
- **Recruiter outreach:** Recruiters frequently use LinkedIn to search for and reach out to potential candidates, increasing your chances of being contacted for a job opportunity.
- **LinkedIn Learning:** LinkedIn Learning offers a huge range of courses for you to upskill. Although this is usually a feature of paid LinkedIn Learning subscriptions, some courses are free (but you may not be able to post them to your profile).

LinkedIn in Languages Other Than English

In China, LinkedIn is known as 领 英 (Lǐngyīng) and operates under a different domain than the global LinkedIn site. However, the functionality and features offered on the Chinese version are largely the same as those on the global site. An alternative is to translate your profile into another language. You'll need to use your knowledge of that language to correct the mis-translations caused by LinkedIn's automation.

The Components of a LinkedIn Profile

There are several components of a LinkedIn profile that you need to include to have an engaging and on-brand professional presentation. Make certain that your photo and cover image, as well as your education and experiences, are set to 'Public'. We've given a brief overview of what to include in each section, but remember that you do not have to include everything.

Think carefully about what the purpose of your LinkedIn profile is. Are you using it primarily for networking purposes or to apply specifically for jobs? This will affect what you put in your profile, particularly your "About" section. Most of the components we've discussed are optional to include.

About

Your 'About' section can be very similar to the Objective Statement you may have included in your resume (see Activity 6.5 to help construct it). It should be short and snappy and adequately set you apart from others. Generally, it is not formed of full sentences because it is obvious that it is about you. Avoid starting this section with 'I am a student studying X at Y university.' This information is freely available on your profile – instead, lead with what you aspire to do, or an achievement that is particularly relevant to the roles you are applying to.

Remember that your LinkedIn profile can be accessed by any recruiter, not just the current job you are applying for. You should therefore try to avoid saying something such as 'My dream job is to work at Google' as this may turn off recruiters who are trying to hire for a different company.

If you are feeling very stuck on your "About" section, try writing a single sentence about each of your resume points. Or use an AI writing program to give you something to start with – you'll need to make it a lot more specific to you. Finally, once you have a LinkedIn profile, you can effectively find and follow other people in your area of interest. Try to emulate (but not word-for-word copy) the language and formatting that they use.

"There's definitely some acquired tricks you learn along the way to get your resume more noticed. Like, for example, sending a LinkedIn message to a person on the team of the role you're applying for. Just saying, hey, I applied to this job and just was hoping you could help me out and see where my status is for this application."

– Lane, chemical engineering graduate and sales engineer in solar PV design.

Featured

If you've done something that you are particularly proud of, such as an award, publication, project or publicly available piece of writing, you can include it here. This section is optional.

Activity

You can't edit anything in this section – it updates automatically when you interact with someone else's post. It is very similar to other social media in that you can 'like' or 'repost' material, and that will show up in this section. When you are actively job hunting, you should be sure to create some activity just by interacting with other individuals. Once a week is a good rule of thumb – when you open your other social media platforms (think Facebook, Instagram, *etc.*), open LinkedIn as well and have a look at whether there is anything useful that you can repost or comment on. Remember to post within business hours for your time zone.

Experience

This section can be copied almost exactly from your resume. One useful feature of this section is that you can 'tag' each experience with skills that will appear as hashtags, and appear in your Skills section near the end of your profile. The recommended number of hashtags per experience is five, but more or less may be appropriate based on the size and scope of the role. Keep in mind that adding these hashtags will make your experience section longer. If there are other valuable things that you want to make sure are noticed on your profile you should consider removing the skill hashtags to combat endless scrolling fatigue.

Education

Again, this section can largely be lifted from your resume (Figure 6.9). Take the time to make sure that you have tagged the higher education institutions that you have studied at correctly so that the icon is correct. These small things can contribute to a cohesive and attractive profile. As per our advice about your resume, do not include your high school.

Figure 6.9   Ensuring that you flag your qualifications clearly can lift your profile; photo by Lewis Keegan on Unsplash.

Licences and Certifications

Activities such as first aid training (physical and mental) or certificates in management can be included in this section, rather than in your Education section. You can also take skills tests on LinkedIn and have them verified here – this is particularly useful if you know a coding language or perhaps are at an expert level in using MS Office. Finally, you can also use LinkedIn Learning to add certifications.

LinkedIn Learning

Most universities and larger organisations have an enterprise licence for LinkedIn Learning – you have access to thousands of free courses in a huge range of areas, from flower arranging to accounting. LinkedIn Learning has taken off in recent years as a valuable way to upskill, often for free or very cheaply. Some organisations ask their employees to complete a certain number of courses each year to keep the employees' skills up. If it's free for you, why wouldn't you use it?

Publications

Highlight any projects or publications that demonstrate your expertise in the field you're applying for. This can include white papers, presentations, or other materials that showcase your skills and knowledge.

Projects

Here are some things you can include in the Projects section:

- **Project name:** Include the name of the project, especially if it's a recognisable or noteworthy project.

- **Description:** Provide a brief description of the project that highlights its purpose, scope, and objectives.
- **Role:** Indicate your role in the project, such as project manager, team lead, or contributor.
- **Skills:** Highlight the skills you used or developed during the project, such as project management, communication, problem-solving, or technical skills.
- **Results:** If applicable, include any quantifiable results or outcomes of the project, such as cost savings, increased revenue, or improved efficiency.
- **Visuals:** Use visuals, such as screenshots or images, to showcase the project and make it more engaging for viewers.
- **Links:** Include any relevant links to the project, such as a website or blog post, that can provide more information or context.

Honours and Awards

This section is quite self-explanatory. It is not necessary to include this section unless you have something unique to brag about. You can tag awards onto the experiences where you developed them as well, to provide better context. Remember that high school honours and awards generally shouldn't be included unless they are showing off a specific length of service in an area that is important to you.

Languages

If you speak or read languages other than English, you can include them here. Make sure that you provide your level of proficiency – 'Native or bilingual proficiency' is the highest level, while 'Elementary proficiency' is the lowest level.

Organisations

The Organisations section is where you can include any relevant memberships that you have. If you are part of a professional group such as the Royal Society of Chemistry, you can showcase that here.

Find the professional body/society related to your discipline. If you are stuck, simply type in your areas of science and the phrase 'professional body'. Or, ask other industry professionals what the important bodies are in their area. Some are niche, so ask them if the ones they know would be relevant for you.

Volunteering

Volunteering is a separate section on LinkedIn. You can include volunteer experience to demonstrate your commitment to your community.

Endorsements

This section can be useful if you have worked on several solo projects or have extensive expertise in a particular area. List your skills and ask for endorsements from colleagues, supervisors, or other professionals who can vouch for your expertise.

Interests: Companies, Groups and Schools

Companies: By following a company on LinkedIn, you can receive updates about new products, services, or initiatives, as well as any major changes or announcements from the company. You may be able to find and connect with employees of the company who can provide valuable insights and information. Additionally, following companies may imply your support for the brand and its values, so be cautious about the companies you choose to follow.

Groups: There are a variety of groups available to join on LinkedIn. Some may also be professional organisations that offer membership. After following groups, you will start to see feeds that can help you learn about the sectors, major trends and challenges (which will feed into your commercial awareness).

Schools: Following schools on LinkedIn is similar to following a company. You can receive updates about upcoming events, news related to the school, and any new programs or initiatives. Some schools offer professional development opportunities, such as workshops or continuing education courses, which may be relevant to your career.

Causes

Select from a predefined list of causes that you are passionate about.

Where to Next?

Now that you have learnt about action verbs, how to construct resumes and cover letters, how to select the best referees and how to examine your documents critically once prepared, it is time to set aside some time to target some specific jobs. If you are still a student, you can use your newfound skills to apply for a better job. Maybe the conditions are better, or maybe the new job you are targeting can add new skills; or choose jobs that will signal a level of professionalism in your resume that earlier jobs do not. If you are a student and genuinely don't want to change jobs (you might have a great job), you can still practice applying. This might mean just preparing the documents and talking with others about how you can improve, but the best test is submitting a few. Best-case scenario, this can lead to interview practice too. Worst-case scenario, you don't get selected for an interview and you can find out why, thus learning more about that company and that industry. You might get to know a few recruitment professionals during this time, too. Knowing what HR professionals are looking for can be a great advantage.

If you are actively looking for a new job, then you are all set to source the best roles to apply for and to work hard on a select number of applications to put in really high-quality applications. Remember this takes time, and it is best to have a plan of how you are going to set aside time each week and how you are going to manage both your excitement and your worry or disappointment if things are slower than you hoped or you do not see success in the time frame you hoped. Be sure to talk to people in your network or read articles to find out the state of the employment market you are applying to. This will help you have realistic aims.

During this time applying for roles, work hard but also be kind to yourself. Missing out on a job simply means there was one other person who the company liked better than you for that specific role. It doesn't mean anything about you personally. And remember that your university or professional group likely has employment help. It may take some looking to find, but these resources can be very important. Every little insight helps.

You have now created a resume and at least one cover letter. You'll need to revisit this chapter each time you apply for a new role to adjust your resume for the specific job requirements. You might use the chapter as a detailed guide initially and then, as you learn the different steps, you might just want to use this chapter to help review fully drafted documents. Keep in mind that when you are applying for a very different position compared to normal, coming back to the basics can also really help you make sure you don't miss a point, which helps you put your best self forward in your application documents.

Remember that you need to make sure your cover letter is specific to the company and role you are applying for, so revising Activities 6.10 and 6.11 is critical. Make sure that your resume and LinkedIn profile tell the same story – you are a graduate ready for the workforce with specific skills that will be of use to your future employer. We reiterate that you need to prepare all of your application documents specifically for each role you apply for. Failing to do this will limit your chances of being offered an interview.

With those final points, we wish you all a rewarding journey through your first few professional job interviews (covered in the next chapter). We both enjoy working in this area because we enjoy helping students and graduates understand the workforce in ways that we never understood when we were in your position. Good luck!

Useful Websites

1. https://jobs.bfftokyo.com/ultimate-guide-to-writing-a-japanese-resume/#The_Japanese_Resume_Rirekisho
2. https://www.hiredchina.com/Chinese-Resume
3. https://nationalcareers.service.gov.uk/careers-advice/cv-sections
4. https://settlement.org/ontario/employment/find-a-job/resume/how-do-i-create-a-canadian-style-resume/
5. https://www.seek.com.au/career-advice/article/what-is-a-resume
6. https://www.gov.za/issues/compiling-curriculum-vitae
7. https://filipiknow.net/resume-sample-philippines/
8. https://resume.naukri.com/resume-samples
9. https://www.resumewriter.my/blog/how-to-write-a-resume/
10. https://www.expatica.com/

References

1. C. G. Berdanier, M. McCall and G. M. Fillenwarth, Characterizing Disciplinarity and Conventions in Engineering Resume Profiles, *IEEE Trans. Prof. Commun.*, 2021, **64**(4), 390–406.
2. J. K. Arnulf, L. Tegner and Ø. Larssen, Impression making by résumé layout: Its impact on the probability of being shortlisted, *Eur. J. Work Organ. Psychol.*, 2010, **19**(2), 221–230.
3. S. P. Joyce, Be Visible on LinkedIn, *Career Plann. Adult Dev. J.*, 2013, **29**(3), 88–90.
4. P. Sterkens, R. Caers, M. De Couck, V. Van Driessche, M. Geamanu and S. Baert, Costly mistakes: Why and when spelling errors in resumes jeopardise interview chances, *PLoS One*, 2023, **18**(4), e0283280. https://doi.org/10.1371/journal.pone.0283280
5. C. Martin-Lacroux and A. Lacroux, Do Employers Forgive Applicants' Bad Spelling in Résumés?, *Bus. Prof. Commun. Q.*, 2017, **80**(3), 321–335.
6. M. Van Toorenburg, J. K. Oostrom and T. V. Pollet, What a difference your e-mail makes: effects of informal e-mail addresses in online resume screening, *Cyberpsychol. Behav. Soc. Networking*, 2015, **18**(3), 135–140.
7. C. S. Diaz, Updating best practices: Applying on-screen reading strategies to résumé writing, *Bus. Commun. Q.*, 2013, **76**(4), 427–445.
8. S. Baert, Facebook profile picture appearance affects recruiters' first hiring decisions, *New Media Soc.*, 2018, **20**(3), 1220–1239.
9. K. Sutherland, K. Freberg and C. Driver, Australian employer perceptions of unprofessional social media behaviour and its impact on graduate employability, *J. Teach. Learn. Grad. Employability*, 2019, **10**(2), 104–121.
10. E. C. Alexander, D. R. Mader and F. H. Mader, Using social media during the hiring process: A comparison between recruiters and job seekers, *J. Global Scholars Mark. Sci.*, 2019, **29**(1), 78–87.

Addressing Key Selection Criteria and Interviews

ROSEMARIE I. HERBERT[a] AND
ANGELA ZIEBELL[b]

[a]Faculty of Science Monash University, Australia;
[b]School of Life and Environmental Science Deakin
University, Australia

If you are working your way sequentially through this book, you will have just developed a cover letter and resume to highlight your skills and your writing ability. However, graduate job advertisements usually ask you to include a statement of Key Selection Criteria (KSC) responses. Then once you have passed the document submission stage of the application, almost every job inevitably includes an interview.

Just as in your resume and cover letter, you'll need to make sure that your KSC responses are tailored to the specific job you are applying for and effectively highlight a range of your qualifications, experience, and skills. For the interview, you'll need to do a bit of research again about the job, company and the interview panel to make your answers (and questions for the panel members) more specific.

We have grouped Key Selection Criteria responses and Interviews because KSC responses and answering behavioural interview questions use the same STARR technique. STARR, or Situation-Task-Action-Result-Reflection/ Relevance is a way to structure your responses logically and concisely to show off your skills. If you have trouble with panicking and not knowing what to say in an interview, having a STARR response you've prepared will help calm you down and make you feel more confident. If you know you tend to ramble, having a STARR response will help to remind you to keep things short and specific. We will start with how to identify the KSC requirements for a job.

Key Selection Criteria (KSC) Responses

Job applications often have KSC to provide a list of specific requirements that the employer is looking for in a candidate. You need to demonstrate how you meet the requirements of the role. In this section, we'll cover why employers ask you to write KSC responses and how to write them. But first, how do you recognise whether a KSC response is required?

What are KSC and Where Do I Find Them?

Key selection criteria are a set of requirements and expectations that employers outline for a job position. These criteria serve as a benchmark against which your qualifications and suitability are assessed. Key selection criteria are typically specific to the job and are designed to match the particular requirements of the position. You must, therefore, tailor your KSC responses every time you apply to a different role.

That being said, there are several common types of selection criteria; (1) Qualifications, (2) Technical skills, and (3) Transferable skills. Addressing technical and educational experience is relatively straightforward, as you can simply write about where you undertook your study, what technical skills you currently have and where you developed them. Instead, we will focus on how to address transferable skills. You should recognise the following transferable skill categories from Chapter 2: Transferable Skills and Reflection.

- Technical competence: Laboratory techniques, research methodologies, or analytical instruments.
- Research and analytical skills.
- Oral and written communication skills: Deliver presentations, write reports or scientific publications.
- Teamwork, collaboration, interpersonal skills, and conflict management.
- Problem-solving and critical thinking.
- Innovation, creativity and adaptability.
- Time management, attention to detail and organisational skills.
- Ability to learn new concepts and adapt to changing circumstances, overcome challenges or learn from failures.

Some job advertisements will have key selection criteria listed under the heading of 'Key Selection Criteria' and mention in the advertisement that you need to address them. Others may include these details under several different headings such as:

- What do I need to succeed?
- Education and Experience
- About you
- The role
- Background, skills and experience
- To be considered for this role, you will have
- Essential skills

Once you have identified the skill requirements of the role, the next thing that you should do is highlight the keywords. When you write your response using the STARR method, you will need to include all of the keywords. For instance, if the criterion mentions "proficiency in molecular biology techniques", the keywords include "molecular biology," "proficiency," and "techniques". Here is another example: "Methodical, with the ability to work in a fast-paced environment". The keywords would be "methodical" and "fast-paced environment".

Although most science roles ask for communication, critical thinking and teamwork skills, they won't necessarily use the same language. In the three sample criteria below, they are all asking about your communication skills. However, they need to be addressed in slightly different ways and with different words. Just as we discussed in Chapter 6: Resumes and Cover Letters, your responses must use exactly the same language as is used in the advertisement, even if the skill is the same. We have highlighted the keywords within each for you.

- Strong **communication** skills to **liaise** effectively with **diverse stakeholders.**
- Excellent **written** and **verbal communication** skills.
- Well-developed **interpersonal** and effective **communication** abilities.

Before you do Activity 7.1 and analyse KSC for jobs you are interested in, take a look at the following example. For this example, there is no KSC section included, but there are skills embedded in the summary of the role. It is unlikely that this company would ask you to write a statement to the selection criteria but you would want to include the relevant skills in your cover letter and resume.

> You will be exposed to varied projects and **work collaboratively** with our clients to **explain environmental approvals**. If you have a **passion** to **continuously learn and grow** as a **consultant** then we want to hear from you.

Addressing Qualification and Technical Skill KSC

The majority of roles you are applying to as a recent graduate will ask for you to have a science degree. Depending on the role, they may also be looking for a particular set of technical skills. Here are two examples of a KSC for a bachelor's degree:

> *A university graduate – in sciences or a technical field.*

> *A completed University Degree in Engineering or Science in the past 2 years or in your final year of completion.*

Activity 7.1: Find and Analyse Key Selection Criteria

Activity goal: To understand how to analyse key selection criteria. Repeat this task each time you look at a new region or a new type of role.

Purpose and benefit: You will have identified key transferable skills that you have in Chapter 2. Use these steps to examine some jobs of interest and identify the skills that your industry is specifically looking for.

Activity steps:

1. Navigate to a job search website/platform such as Indeed or LinkedIn. Revisit Activity 6.1 to help you examine the parameters that you can choose to find jobs that specifically interest you.

2. Pick three jobs that sound interesting and open each in a new window or tab (or print them) and review the following sections.
   - Job description or summary
   - Your responsibilities | About the role | Key accountabilities
   - About you | Selection criteria | To be successful

3. Save the files, then open your saved files with annotation software (or take your paper) and use the highlight and comment tool to annotate the KSC with the broad categories for the technical and transferable skills required for the job. If there is no KSC section, look at the job description and see what skills are mentioned within it.

4. Identify and highlight the keywords within each criterion. Are these similar in each job advertisement? What patterns do you see in the keywords? You should notice that there is more than one way that you can be asked about a single skill.

5. Remember that if you don't meet all the skills, you should make a plan to improve that skill over the next few months. What skill will you work to improve? How?

Takeaway: You will now be able to analyse a job advertisement to identify the KSC within the advertisement and identify the keywords that you should use in your response. You might have identified some areas where your skills are insufficient – flip back to Chapter 2 for some ideas on how to improve them.

To address these, you could have a simple statement such as:

> *I hold a Bachelor of Science from [X University] and graduated in [month, year]. I majored in [subject] and minored in [subject]. I developed technical skills in qPCR, HPLC and modelling using GraphPad through my studies and in a 6-week research project.*

Now that you know what to look for and what to address, we will cover which kinds of organisations ask for a KSC response and how long each of your responses should be. We will also have a quick revisit of formatting guidelines.

Who Uses KSC and Why?

There is a tendency for larger companies to ask for KSC responses because it enables them to sort large volumes of candidates into a short-list for interviewing. KSC responses can provide a clearer assessment of a candidate's skills and abilities than resumes and cover letters. KSC responses are designed to measure your ability to think critically and communicate effectively. It enables a non-specialist hiring manager to judge candidates based on the match to the KSC asked for. KSC responses can also be scored quickly and easily by a computer, which saves employers time and money. This makes KSC responses an attractive option for large companies that are constantly looking for new talent. Companies that often ask for KSC responses are large pharmaceutical and biotechnology companies, government agencies, regulatory bodies, healthcare providers, and medical research organisations.

Smaller companies are less likely to have key selection criteria for you to address. KSC requirements can be time-consuming, both to write the criteria to be included in the job advertisement, and to read responses and rank applicants. Smaller companies also usually have a more informal hiring process, with fewer layers of review and decision-making. Smaller companies often place a significant emphasis on cultural fit as well. If you aren't asked to write a KSC statement, you should make doubly sure that your cover letter and resume illustrate your experiences and demonstrate your ability to write clearly and persuasively.

How Long Should a KSC Response Be?

You'll need to follow any page or word limits provided by the employer or application instructions. Some companies, particularly government roles, will provide a supplementary document on how to address their particular application. Adhering to these guidelines demonstrates your ability to follow instructions. If no guidelines are given, the length of your response will vary depending on your level of relevant experience to a specific criterion.

As a recent graduate, you might not have much (or any) relevant experience. The response should be succinct; typically, around 200 to 300 words per criterion. You just need to have one really good example to address the KSC. We have addressed what to do if you don't have relevant experience in section "Addressing KSC You Do Not Meet". Keep your responses concise and to the point. Do not use bullet points because this is an opportunity to show off your writing skills.

> *"I think you need to when you're applying for jobs really focus on other qualities that you have as well and what else you can bring to the table, because you're a whole person not a number on a piece of paper."*
>
> – Ruth, science graduate working in the retail food industry as a line manager, former business owner.

Once you have extensive experience in the field, aim for a response length of around 300 to 500 words per criterion, depending on the application guidelines. Sticking to using the STARR method will help you avoid the temptation to include unnecessary details or lengthy explanations. Focus on the most relevant and impactful aspects of your experience and qualifications.

Before submitting your application, proofread your responses to ensure that they are free of errors and are well-structured. Also, a quick reminder that your formatting does not need to be particularly fancy – it just needs to be functional. It should be practical and easy to look through, with clear headings for each selection criteria. Remember the file naming conventions we taught you in Chapter 6. Give your KSC responses (and resume and cover letter) a sensible filename such as "Lastname-KSC-jobnumber".

> **Hint:** AI text generators such as Gemini and ChatGPT write terrible KSC responses. This is because a good response needs to give a specific example of when you have used a skill – AI does not know about any of your experiences or achievements. It also tends to generate verbose, generic and formulaic responses that are easy to detect as written by AI, even when they follow the STARR format.

Addressing KSC: The STARR Technique

STARR is an acronym that stands for Situation, Task, Action, Result, and Reflection/Relevance. It is a widely used framework for structuring and providing evidence-based responses to KSC in job applications or interviews. The STARR method helps you to convey your skills effectively with evidence from your experiences.

Situation: Describe the context or circumstances in which a particular challenge or task was encountered. This sets the stage for the rest of your answer. It can be very helpful to include quantitative information such as "I managed a team of six people who delivered handwritten, individualised shipping for more than 1000 online shoppers per day".

Task: Identify the goal or objective that needed to be achieved in that situation. What is the goal you needed to achieve? "I needed to prioritise packing and mailing based on the distance and shipping time to ensure items were delivered before Christmas".

Action: Detail the specific steps taken to address the situation or complete the task. Focus on your personal contribution and highlight the skills or strategies you used. Use action verbs.

Result: Explain the outcome of the action taken and how the situation was resolved or the task was completed. Highlight the positive impact you made, such as resolving issues, improving processes, or delivering successful outcomes. Quantitative results or specific achievements can be particularly useful.

The final R refers to either Relevance or Reflection. Relevance can be used in a KSC response to show that you have specifically created your statement to meet the requirements of this company. Reflection is useful in a job interview where you can illustrate your growth mindset and ability to learn from mistakes.

Relevance: Explain how the skills or qualities demonstrated in your example directly align with the criteria. Connect the dots between your actions and the specific skills or attributes the employer is looking for by including the company name and/or specific values.

Reflection: Reflect on the experience and its significance. Discuss what you learned from the situation and/or how it has contributed to your professional growth. This demonstrates your ability to analyse situations, learn from experiences, and apply knowledge in future contexts.

Composing a STARR response can be quite difficult when you first start writing them. It is helpful to look at a range of KSC responses to help you start writing and refining your own. Both of the following examples address "The ability to work in a fast-paced environment" but Example 1 is not concise (240 words) and is vague about what the candidate achieves in their role. It is also quite repetitive and uses many synonyms – and doesn't include exactly the same keywords as given in the KSC. Example 2 gives a specific situation that the candidate resolved in 120 words and matches the keywords in the KSC. The first couple of times that you write a STARR example, it will probably turn out like Example 1. Do not worry, you will improve as you practise and seek feedback and examples from others in your environment.

Example 1. *Working as a server in hospitality has helped me to develop my ability to work quickly. I have had to juggle several jobs/tasks at once and identify the tasks that require immediate attention. While some were easy to handle, some required me to be in two places at the same time. This includes taking orders, serving customers, managing payments, and ensuring the overall smooth functioning of the café. The diverse range of responsibilities has challenged me to stay organised and make quick decisions on the spot. In a bustling café, numerous demands are vying for attention simultaneously. By looking at the urgency and importance of each task, I have become good at prioritising and tackling them in a logical order. I have the ability to assess the urgency of each situation, delegate tasks to my colleagues when necessary, and handle multiple demands with composure and efficiency. I had to also understand and resolve the problems my other staff members were encountering in a professional manner. I have proactively assisted my colleagues when they encountered challenges, offering solutions and support to ensure smooth operations. I have often been commended by my co-workers for my ability to work through high-tempo environments. In summary, my experience as a server in a busy café has provided me with invaluable skills in working quickly and efficiently. I have mastered the art of multitasking, prioritising tasks, resolving problems, and maintaining composure in high-pressure situations.*

Example 2. *As a manager at [restaurant] I have developed my ability to work in a fast-paced environment. I coordinated a team of five staff to deliver an end-of-year staff function of around 100–200 people. On one occasion, several staff members did not show up for their shifts. I prioritised by identifying the most critical tasks and assigning tasks based on skills and experience. I ensured that everyone knew which tables they were responsible for. This meant that customers received smooth and consistent service throughout the evening. We then received a positive Google review from the organisation that had organised the event. In the role of laboratory manager at [your company] I will maintain high performance standards while working in a fast-paced environment.*

A common error that often leads to a too-long answer is to write a lot about the Situation and Task and neglect to write a detailed account of the Actions you took. Theoretically, anyone who worked in the same job as you will be exposed to the same Situation and Task, but your personal Actions are what the hirer wants to hear about. Asking someone else to read over your STARR response will help you identify if you've made this mistake.

In Activity 7.2, we have given you a generic example of a STARR response to give you a framework to follow. You'll need to individualise it by including a situation that you found yourself in and a clear result. The main way

Activity 7.2: Improve a STARR Response to a Generic Transferable Skill KSC

Activity goal: To brainstorm your relevant experiences and use them to write a convincing STARR response to a generic KSC.

Purpose and benefit: Writing a KSC response using a template will make it easier for you to create your own responses to specific KSC. Use these steps to familiarise yourself with using the STARR framework.

Activity steps:

1. Analyse the following KSC response that has been written in response to the skill "Communication with diverse stakeholders". Analyse whether it follows the STARR framework and incorporates the keywords from the advertisement.

> Communication is a vital skill in any professional setting, and my experience has taught me the importance of clear and effective communication. I have gotten good communication skills through various experiences, such as working in teams, dealing with customers, and collaborating with different people. I have to change my communication style to suit the audience so that everyone is satisfied with the customer service they get.

2. You should have noted that this is a very vague answer that states the same information multiple times and doesn't give a specific example. It says that communication is important, but doesn't actually say what is being communicated. It is not clear where these skills were developed, how big the teams were and what type of stakeholders were involved. It is also missing action verbs.

3. Brainstorm a couple of situations where you have communicated with diverse stakeholders. You might have a professional example where you worked with managers, social media coordinators and the general public. Or, if you have tutored a student, you may have communicated with the student, their parents and their regular teacher. Now, select the example that best demonstrates your communication skills.

4. Use the STARR method to structure your response, incorporating the example that you have chosen. Make sure that your example is specific, and highlights your strengths. Remember to put very little emphasis on the Situation and Task, and more on the Action and Result. You won't be able to include the Relevance in your example because you aren't addressing a specific job.

5. Check how long your response is – remember to aim for 250 words. Have you used action verbs? See the Chapter 6 section 'Action Verbs, Dynamic Verbs or Strong Verbs – Illustrating Your Skills' for a refresher on how to use them. Have you named the stakeholders and described your communication style? Is the result/outcome clear?

Takeaway: After looking at a poor example of a KSC response, you should then have written one for yourself. Communication is a very common KSC skill, so now you'll have a response ready to modify for specific jobs.

to improve the quality of your STARR response is to write many, many responses. Asking your mentor to help you by providing examples of good KSC for the environment you are applying to is another great way to improve your responses.

You've now had the opportunity to re-write a generic KSC response, but it is not written for a specific job. Remember that each time you apply for a job, you need to personalise your KSC responses to make sure that your language matches that used in the advertisement. Once you have identified the keywords in the criteria itself, you should then revisit the rest of the job advertisement. If there is text further describing the day-to-day responsibilities of the position, you can use this to help pick further keywords to include in your response.

Here is an example. This advertisement's KSC asks you to address the following criteria:

- *strong decision-making and communication skills,*
- *resilience, enthusiasm and a passion for teamwork to achieve positive outcomes in a deadline-driven environment.*

When you check the text of the advertisement, it also includes this information:

You will be responsible for promoting compliance and excellence in all environmental obligations and positively influencing project stakeholders to achieve the best project environmental outcomes.

Here is an example of a 250-word STARR response written to address the first criteria. We have bolded the keywords used to make it easier for you to see how the language matches the criteria. Keep in mind that you wouldn't bold words in a real example submitted to a hiring company, and you also wouldn't include the section titles of Situation, Task, *etc.*

Situation: As an intern at [Environmental Solutions], I used my ***decision-making skills*** during a large-scale construction project that focused on ***environmental compliance*** and ***stakeholder*** engagement.

Task: My role was to clearly communicate to ***positively influence project stakeholders*** and thus ensure adherence to environmental regulations.

Action: I evaluated the environmental implications of a new private roadway in a housing estate on a nearby wetland. I assessed potential risks and impacts and ***made decisions*** on how to best minimise ecological disturbances and protect sensitive habitats. I presented the environmental management plan and discussed potential impacts. I tailored my ***communication*** to different stakeholders. For construction workers, I focused on the practical aspects of environmental compliance and the positive impact it would have on their working conditions. When I engaged with senior management and regulatory authorities, I emphasised the legal and reputational benefits.

Result: As a result of my **strong decision-making and communication skills**, the project received positive feedback from stakeholders and met environmental compliance measures. I **positively influenced project stakeholders**, fostering a shared commitment to achieving the best environmental outcomes for the project. Moreover, my proactive approach and clear **communication** contributed to improved project efficiency and cost savings.

Relevance: As an employee at Environmental Change Pty Ltd I will be able to **communicate** confidently with diverse stakeholders and use my **decision-making** skills to work independently where required.

As you've seen in the example above, you need to weave the KSC words and text from the advertisement with your example. While you may be able to reuse parts of your key selection criteria response for each job you apply for, you must match the language and wording exactly (see the Chapter 6 section 'Applicant Tracking Systems' to understand how applicant tracking systems assess your application documents). The final 'R' of Relevance is specific to the company and emphasises that you've spent time customising your response for a specific role. In Activity 7.3, we will walk you through addressing a specific KSC using the STARR technique.

Alternatives to STARR

We hope that you are now confident using the STARR response to structure your KSC responses. However, not all experiences you have will fit nicely into the STARR acronym, as perhaps the situation and task are difficult to separate. You can use CARR or PARR instead.

CARR (Challenge, Action, Result, Relevance/Reflection): This method is similar to the STARR method, but includes a focus on the challenge or problem that you encountered. You start by describing the challenge or problem, then move on to explain the specific actions you took to address it, and finally detail the results that were achieved as a result of your actions.

PARR (Problem, Action, Result, Relevance/Reflection): This method is similar to the CARR method, but focuses more specifically on the problem that you encountered. You start by describing the problem, then move on to explain the specific actions you took to address it, and finally detail the results that were achieved as a result of your actions.

Activity 7.3: Craft a STARR-structured Response to a Specific KSC From a Job Advertisement of Your Choice

Activity goal: To write a specific KSC response using the STARR framework to present your relevant experience. Repeat until you have completed an analysis for all the KSC you are seeing in roles you are interested in, or you feel that you have learnt the technique very well (refer here occasionally to check in).

Purpose and benefit: You likely have not written a KSC response before, or if you did, it may not have followed the STARR format. Use the steps in this activity to help you write one that is targeted to a KSC in a job you are interested in.

Activity steps:

1. Find a job advertisement that looks interesting and has a clear KSC section for you to address.

2. Choose one of the key selection criteria from the job advertisement or position description then analyse and annotate the keywords that you will need to include in your response.

3. Examine the rest of the job advertisement to find other keywords that are relevant to include. You will need to research the job requirements, the company's values or mission, or other relevant information.

4. You should have already brainstormed your examples to use for a generic KSC response. You may have created a list or other visual aid – now you should generate ideas for examples that demonstrate how you meet each of the specific components or requirements of the KSC. These examples should ideally be different for each KSC that you address.

5. Use the STARR method to structure your response, incorporating the example that you have chosen. Use at least two action verbs to make your example seem more active and as if you are in control of the events. You should include the second 'R' at this stage – how are the experience and skills you are covering in the response relevant to the job you are applying for?

6. Review your response and refine it to ensure that it meets the specific requirements of the KSC and is tailored to the job listing. The language also needs to exactly match the KSC given in the advertisement.

7. Now, check for spelling and grammatical errors. If possible, seek feedback from a mentor, friend, or colleague to get their input on your response. Use their feedback to refine your response further.

Takeaway: Following these steps will help you craft a strong KSC response that demonstrates your skills and experience. You will want to repeat this activity each time that you apply for a job to make sure that you are addressing the KSC that are given. Remember that each different KSC statement needs a different paragraph to address it. Try to use a different experience for each response, *e.g.*, one from university, one from a part-time job, one from a volunteering role.

The key is to use a method that works for you and allows you to provide clear, concise, and specific examples of your skills and experience.

Addressing KSC You Do Not Meet

As we discussed in Chapter 6, it is likely that you will not fit all the criteria. Job descriptions are an employer's wish list – they want to recruit talented employees who have as many skills as possible and fit well into their team. They can't expect you to know everything, and they realise that they may need to train you in some skills. It is your job to convince them that, despite not having the skill, you are a fast learner and will be able to pick it up quickly on the job. If it is a skill that is common to many of the jobs you are applying to, you should plan to improve or gain this skill independently.

If you do not meet a KSC, start your response by explaining why you do not meet it. Highlight any similar skills you might have. For example, if the advertisement has asked you to be able to code in Python, you could instead say that you have skills in Perl and therefore can pick up a new coding language quite quickly. Then, explain how you are planning to gain/improve the skill by taking courses or independent study.

Here is an example of how you could address a KSC if you don't meet it:

> *I understand that this position requires experience using C++. I am not yet confident in coding in C++, but I have three years of experience using R and Java. I am undertaking LinkedIn Learning courses to improve my C++ knowledge and I am confident that I can quickly become proficient in C++.*

What's Next?

Writing KSC STARR responses will be time-consuming until you have a larger bank of answers to modify critically and personalise for each job that you apply for. Remember that in every application, you need to make sure you have incorporated any relevant information from the job advertisement in your response, as well as using exactly the same keywords as listed in the KSC. The second R in STARR will be particularly helpful.

In the next section, you will use the STARR method to address questions in an interview. You'll need to phrase the KSC in a job advertisement as a question, then write a response that follows the STARR method. This will help you to keep your responses concise and to the point.

Prepare to Be Interviewed

Congratulations. You've reached the all-important interview stage of the job application process. You might be dreading it – do not worry, we have plenty of great advice to help set you up for success. We will cover how to prepare for an interview, from choosing the right clothes to wear to following up after a successful interview. Improving your interviewing skills is about practice and having prepared answers to potential questions. You will reuse the STARR technique to address interview questions, which is why we have linked KSC responses and interviewing together.

"Your anxiety [before] a job interview is a function of your preparedness. If you're not prepared, if you haven't prepared, if you haven't looked at relevant questions and drafted a short response to give you something to say, you're going to feel a lot worse than if you have taken the time and familiarised yourself with the possible things that can come up."

– Dylan, science/arts double degree graduate with a psychology major, employed by a retirement fund company.

Some people feel that they'll do naturally well in an interview because they are very confident about themselves. However, it is not just about your confidence in yourself – you need to be able to articulate your skills and interest in the company. Do not rely on the adrenaline of the experience to boost your confidence going into the interview. You can certainly approach the interview with confidence and a positive attitude, but being prepared is going to help you feel more comfortable, and therefore confident.

"We don't rise to the level of our expectations, we fall to the level of our training."

– Archilochus.

"I don't think confidence or self-esteem is a prerequisite for a good job interview. I think that practice is."

– Dylan, science/arts double degree graduate with a psychology major, employed by a retirement fund company.

How to Schedule an Interview

There are several ways in which a company might schedule an interview with you. You might be sent an email with several times you can select from. If there are a lot of applicants for the job, you may just be told about one time when you can do the interview. If you have a choice, aim to schedule your interview in the middle of the week, or at least not on Monday morning or Friday afternoon. Also do not pick a time that is too close to lunch hour, as studies have shown that this time can negatively impact the ability to focus for yourself and your interviewer.[1]

Alternatively, the hiring manager may call you. If you are unable to answer a phone call due to being in a noisy environment or similar, it is better to let it ring through to voicemail and ring them back at a time when you are composed. If you have a voicemail option, make sure that your answering message is polite, *e.g., Thank you for calling [name]. I'm not available at the moment, so please leave a message with your name, phone number and reason for calling.*

Answering Questions in an Interview

The reason we have grouped answering key selection criteria and interviews in one chapter is because you should use the STARR method to structure your answers for both. With an interview, you will likely be asked specific questions that are designed to encourage you to talk about your past experiences and how they make you a good candidate. If you've ever felt like you weren't sure what to say in an interview (or, conversely, if you talk too much), having a STARR response ready will help you to feel more confident.

To prepare for an interview, you should reread the job description carefully to ensure that you understand the required skills and responsibilities of the position. Make sure that you give yourself plenty of time to prepare comfortably (Figure 7.1). If the job description includes a Key

Figure 7.1   Put aside time to schedule and prepare for your interview thoroughly without rushing, which can be stressful. Photo by Bruce Mars on Unsplash.

Selection Criteria or a similar list of skills and attributes they expect you to have, this list will help you identify potential questions to answer. Here are two examples of common KSC: "Time management, organisational and problem-solving skills" and "Excellent verbal and written communication skills". Then, we've framed them as questions:

- Are you an organised person?
- What is your approach to problem-solving?
- How would you describe your communication style?

The three examples we have just given you are quite straightforward. Most interview questions are more complicated and require you to tell a clearer story. The next two we've included are a little trickier – your answer needs to cover more than just problem-solving; it should also include reference to your communication and/or interpersonal skills.

- Tell me about a time when you experienced an unexpected challenge in the workplace.
- Describe a situation in which you found a creative way to overcome an obstacle.

There are often different levels of questions for the same topic, *e.g.*, Tell us about the last team project you worked on. Tell us about when you needed to deal with conflict within a team. Tell us about a time when you worked as the team leader. Work examples are usually more powerful than university examples for proving team skills because you have an ongoing relationship with those workmates (university project groups usually only last for a semester or less).

Here is a selection of sentence starters to help you compose your own questions in Activity 7.4.

- "Tell me about a time when..."
- "Give me an example of..."
- "What was the most difficult situation you've had..."
- "Tell me about a disagreement you've had..."

Should I Talk About My Expected Salary In the Interview?

It is generally considered bad manners to talk explicitly about salary expectations in an interview because the primary purpose of an interview is to assess your qualifications, skills and experiences, and your fit for the role and organisation. Asking about salary expectations in an interview can shift the focus away from your fit and lead to potential biases in the evaluation process (*e.g.*, hiring someone who is 'cheaper'). Some candidates may also be at a disadvantage and unintentionally disclose a lower salary expectation.

Activity 7.4: Create an Answer Bank for Potential Interview Questions

Activity goal: Create an answer bank for potential interview questions. Repeat for each job, while this might seem excessive with related jobs, it is important to put in the effort. Related jobs will be faster to prepare for.

Purpose and benefit: By having an answer to each question related to a KSC you will be able to flexibly and naturally answer questions that are relevant to the role you are applying for.

Activity steps:

1. Navigate to a job search website (*e.g.*, Indeed or LinkedIn). Search for job titles that you are interested in. Pick one of these job listings to address – remember that if you can't see a set of clear requirements for the role, there may be a PDF document with more details on the web page. Refer to section "What Are KSC and Where Do I Find Them?" for potential other titles for KSC to help you.

2. From the requirements of the role, identify and annotate what technical and transferable skills are needed, *e.g.*, communication, problem-solving, or teamwork abilities. Refer back to Chapter 2 if you have difficulty identifying these skills.

3. Examine the rest of the job advertisement and research the company's values or mission.

4. You should have already written some KSC responses in Activities 7.2 and 7.3. Pick a couple of examples that represent a range of your experiences and use these to get started.

5. Use the STARR method to structure your response and use bullet points and short phrases. You should include the second 'R' of Reflection – how have you further improved your skill since the experience, or Relevance – how are your skills related to the job you are interviewing for?

6. Review your response and refine it to make sure it is concise, with less emphasis on the Situation and Task. Use your bullet points to create sentences as you read it out loud. You need to avoid just memorising your response word-for-word as this can cause you to sound monotone and boring. You are trying to tell a story that creates a picture of yourself in the workplace.

7. Practice your responses out loud to see what they sound like. You should do this in front of a mirror or as a video recording if possible. Watch back your answer – does it seem genuine? Or do you sound quite robotic?

Continued

Activity 7.4: Create an Answer Bank for Potential Interview Questions – (*Continued*)

8. Seek feedback from a mentor, friend, or colleague to get their input on your response. Use their feedback to refine your response further. When you are feeling confident, do a full practice interview rehearsal with someone else (we will guide you through this in Activity 7.5).

Takeaway: By following these steps, you can develop a bank of questions and answers for a specific job interview. Over time, you will develop a range of answers that you can refresh your mind about before each interview. Do not get too confident though. Remember that every company, even within the same industry, is going to be slightly different. You are still going to need to educate yourself about the company and make sure that your answers address their needs.

Before you go into an interview, you should have gotten a good grasp of the salary range of the roles you are applying to. Then, if you are forced to name a salary during the interview, you can at least give a range. Asking about salary expectations allows interviewers to gather valuable information about the current market rates and salary expectations within the industry. This information can help the company remain competitive in attracting and retaining top talent.

Inquiring about salary expectations too early can create the perception that the company prioritises compensation over other important factors, such as job satisfaction, growth opportunities, company culture, and alignment with the company's values.

Instead, the interview can be a good place to ask about other benefits of the role, such as mentoring programs, paid time off, *etc*. In Activity 7.6, we will help you to put together a list of potential questions to ask your interviewer.

"I was applying for everything for the sake of it. Part of me actually really wanted to just get practice doing job interviews. That was something that I was poorer at. I could write a really good resume, I could write a really good cover letter. But in the pressure of that job interview situation, I was not as confident. My motto is just do it anyway."

– Dylan, science/arts double degree graduate with a psychology major, employed by a retirement fund company.

Practice answering questions out loud, maintaining good eye contact, and projecting confidence. Consider recording yourself to assess your body language, tone of voice, and clarity of expression.

You also need to research the company to learn more about its history, mission, values, products or services, and recent news. This will help you to understand the company's culture and goals, and will also help you to tailor your responses during the interview. It will also give you a better idea of how you will fit into the company. Revisit Chapter 3: Commercial Awareness for more details about finding and understanding commercially relevant information to help you prepare for applying for a job. Remember to prepare specific examples to showcase your commercial awareness and how it aligns with the company's goals and values.

In your invitation email, you should have been told who is going to be on your interview panel. Use LinkedIn or the company's social media and web page to research each of the people you will meet. Knowing what they look like will help you feel at ease when you get there and are facing them (Figure 7.2). You will also know what aspects of the job and associated skills they might prioritise.

Remember that if you are hired, you might work with or under the guidance of the people on the panel. As much as they are interviewing you for the role, you should also think about whether you can work with them productively. Treat the interview as an opportunity to learn more about the role.

Figure 7.2   If you know something about the people in your interview panel before the interview starts, you will better be able to answer questions and make conversation with them. Photo by Van Tay Media on Unsplash.

You may have access to a virtual job interview platform as part of your university careers service. Depending on how advanced it is, you may be able to get feedback on your answer, tone and posture. You can also use an AI text generator to generate interview questions for you and simulate an interview in that way. Running a practice interview with a trusted friend or family member can be an excellent way to prepare for an upcoming job interview. Activity 7.5 will guide you through the process of conducting a practice interview effectively.

Activity 7.5: Practise Your Interviewing Technique

Activity goal: To familiarise yourself with the interview process to increase your confidence. This task will take 1–1.5 hours. Repeat a few times for different jobs to build confidence, also repeat when the type of role is different.

Purpose and benefit: Develop your ability to answer interview questions coherently by using feedback from yourself and others.

Activity steps:

1. Choose a job listing of interest to practise interviewing for. Ideally, the position should have some key selection criteria to help you come up with some interview questions.

2. Create a bank of around 10 questions based on the role. You should have at least two to three questions each about technical skills, transferable skills and behavioural/situational questions.

3. Research the company and reflect on your experiences that can be used to address the questions you have developed. Prepare at least one STARR response for each question, and incorporate your knowledge about the company and industry into your answers.

4. Explain to your friend or family member about the job position you're interviewing for, including the company and any specific information that might be relevant. Provide them with your list of interview questions or ask them to create their own questions based on their understanding of the role. Ask them to take notes on your performance – what went well and areas where you can further improve.

5. Ask your interviewer to start the interview by asking "Tell me a little about yourself". They can then ask any of the questions that you have prepared in step 2. Answer the questions as if you were in a real interview, using the STARR method.

Activity 7.5: Practise Your Interviewing Technique – (*Continued*)

6. Wrap up the interview by asking a question of the interviewer (prepared in Activity 7.6). They can just make up an answer from their own experience.

7. Your interviewer can provide feedback on your answers, including strengths and areas for improvement. Take notes of any additional points you want to remember or need to think about improving.

8. After the practice interview, take a moment to reflect on your performance and the feedback you received. Did you feel more comfortable while interviewing? When you answered the questions, did you experience a mind-blank and forget what you were going to say? Refine your responses if needed – were they too long or too short? Did you ramble?

9. You can repeat this activity as many times as you like. Running multiple practice interviews with different sets of questions and/or different interviewers will build your confidence.

Takeaway: Interviewing is something you can improve, and practising can help you become more comfortable, articulate, and confident. It allows you to receive valuable feedback and make necessary adjustments before the actual job interview. Remember, practice and preparation are key to performing well in interviews.

"There's three different topics that they hire towards and they weigh them appropriately. Do they like you and do you have aptitude for the role? Would they enjoy working with you? And then three, like, do they see growth potential in you as a person?... The aptitude is 10% to 20% of how much experience they have in the role, and then it's like 30 or 40% for do they want to work with you and 30 or 40% do they see you as a person who can grow with them in the company?"

– Lane, chemical engineering graduate and sales engineer in solar PV design.

On the Day of the Interview

Before the Interview

Plan What to Wear

Knowing what to wear for an interview is important to make a positive first impression. Wearing clothes that make you feel more confident will

Figure 7.3   What to wear is important but all you need to do is follow the basics and try not to stress too much about it. Photo by Ruthson Zimmerman on Unsplash.

also give you a boost to go into an interview knowing that you are presenting your best self (Figure 7.3). When you are invited for an interview, you should do some further research to find out what to wear. Revisit Chapter 4: Networking for some ideas on how to dress.

Research the company dress code to understand the company's culture and expectations. Check the company website or social media pages for photos of employees to get a sense of what is appropriate attire. When in doubt, dress one level up from the company's dress code. For example, if the company dress code is business casual, consider wearing business professional attire for the interview. Remember that it is always better to be overdressed than underdressed for an interview. As a new graduate, you may not have a large budget for buying clothes – remember that it is very unlikely that a full three-piece suit is required. Buying from second-hand shops gives you a chance to try out a range of clothing types that may help you feel more comfortable. You should consult a trusted professional contact on whether what you have chosen is appropriate.

You should choose conservative attire that is clean, pressed, and well-fitted. Try to avoid wearing anything too flashy or distracting – you want to be judged on your answers to the interview questions, not what you are wearing. A way to do this is to wear neutral colours such as black, grey, navy, or white. Make sure that you have comfortable, appropriate shoes for the role. Do not wear sneakers, flip-flops, or any shoes that are too casual.

Eat Something

Having something in your stomach before the interview will help in a couple of ways. You'll be better able to focus on the questions being asked rather than on your rumbly stomach. Eating anything will calm your flight/fight/freeze instincts, so you will feel less nervous. If you usually have a coffee or caffeinated drink in the morning, don't skip it as you'll likely develop a caffeine-withdrawal headache or feel a bit cranky before your interview. Be aware that if you don't normally drink coffee, it could be a bad idea to indulge, as it could lead to gastrointestinal upset.

Being well-fed can help you project confidence, enthusiasm, and a positive attitude during the interview. It is important to choose foods that provide sustained energy (not a sugar high) and avoid heavy or greasy foods that may make you feel sluggish. Remember to drink enough water to stay hydrated as well, but don't drink more than a glass or two in the hour before the interview, or you may find yourself needing to go to the bathroom during the interview.

Be on Time

Plan to arrive at least 10–15 minutes early for the interview to allow time for any unforeseen circumstances and to avoid rushing. If you are worried about the trip, you can do a practice run on the day before the interview. This can also be a good thing to do before your first day of work as well. Arriving late to the interview will potentially give the panel the impression that you either do not care about the interview (or by extension the job), or that you often run late and are not reliable (Figure 7.4).

Figure 7.4 It can be incredibly stressful before an interview. Don't risk being late and adding to that stress. Photo by Musa Haef on Unsplash.

If you arrive more than 15 minutes early, find a quiet place to sit and read over your prepared notes – you probably do not want to run into the previous candidate interviewing.

Take a moment to compose yourself – take a few deep breaths and calm any nerves you might have. Make yourself known to the receptionist or check-in person and let them know that you have arrived for your interview. Treat everyone you encounter with respect and kindness, from the receptionist to any other employees you may encounter. You never know who may be evaluating you.

You may be asked to wait in a lobby or waiting area until your interviewer is ready for you. Use this time to review your resume and notes, or to practise your interview responses. Make sure your phone is switched off. While you are waiting, pay attention to the surroundings and company culture. This can help you get a sense of what it might be like to work there and tailor your responses accordingly.

If You Are Unable to Attend the Interview

Occasionally, a situation might arise and you are unable to attend your interview. You might be ill, in an accident or need to care for a family member at the last moment. Make it a priority that you call the phone number in your interview invitation and let them know that you can't attend. If you are a top candidate for the role, they may be able to offer you an alternative time to be interviewed. Remember that industries can be quite small – you do not want to be known as the interviewee who stood everyone up and wasted their time.

Alternatively, maybe you have just gotten a job offer somewhere else and intend to take it. Remember the advice of our recent graduates – there is no harm in going to the interview for practice. Additionally, you might seek employment at the company at a later stage in your career, so do not burn any bridges by rudely not turning up.

During the Interview

Smile, make eye contact, and introduce yourself to the panel. Shake hands if appropriate – pay attention to their body language for whether a handshake is expected. Take a deep breath, relax, and be yourself. Usually, the chair of the panel will tell you a little more about the role, and then give a summary of how the interview will be conducted. Remember that some of the panel might not have looked over your job application documents, and sometimes panel members may have been brought in at the last minute to interview you.

Hint: If you are feeling particularly anxious, a good distraction can be to wiggle your toes to let off some nervous energy. Other simple ways to decrease your nervousness are to focus on each of your senses in turn and take some deep, slow breaths.

A very common first question is "Tell us a little about yourself". This question aims to break the ice and get to know you better as a person, beyond your resume or cover letter. Everyone gets nervous at an interview, and interviewers know this – so this question is helpful to set yourself up for success. Be prepared for this question and have a concise answer ready. The interviewer wants to see how you present yourself and communicate and also wants to get a sense of your personality and interests. The interviewer is also trying to assess if you are a good fit for the company culture and the position you are applying for. Your answer to this question can set the tone for the rest of the interview.

Instead of fixating on the outcome of the interview, shift your focus to the present moment and the conversation at hand. Engage actively with the interviewer and listen attentively. By directing your attention to the discussion itself rather than worrying about the future, you can reduce anxiety. Think of the interview as an opportunity to learn more about the company, even if you don't get the job.

There is nothing wrong with asking an interviewer to repeat the question or say that you do not quite understand what they are asking. Doing this may indicate to the interviewer that you are willing to ask questions to avoid getting things wrong, which could make you more likely to get the job. Not knowing how to do absolutely everything and asking for help when necessary is an attractive quality in a potential employee.

"I realised that the more informal the interview was, the more successful it ended up being because they're like, okay, cool. This person obviously did some homework. So they check that box. And the rest is, do I think that they would jive with the team? And do I think that they will be a value add to me enjoying my job? And then it's like, okay, do I think I can train this guy up on the things I don't know?"

– Lane, chemical engineering graduate and sales engineer in solar PV design.

A longer interview is better for both you and the hiring company. You get more of a chance to break through subjective filters and demonstrate why you'd be a good candidate. The interviewer gets more information, and bases less of their decision on immediate, emotive impressions. That being said, interviewers can form their opinions very quickly in the interview, so it is important to make the right impression from the very start.

Asking Questions in an Interview

At the end of an interview, interviewers will almost always ask whether you have any questions. To some extent, you can predict what they are expecting you to ask about – possibly more details about the position, working hours or perks. You'll need to have a bank of around four to five questions to ask, just in case the question you were going to ask was covered by the interview already. There is no such thing as a 'stupid question', but you want to avoid repeating yourself at the risk of looking like you weren't paying attention while hearing about the role. Declining the opportunity to ask questions can indicate that you are uninterested in the job. Conversely, the questions you ask can be used to subtly reinforce your suitability for the role.

When you write the questions, make sure that they are open questions – questions that do not have a yes/no answer. The question "Are flexible hours available for this position?" might result in a short answer. Instead, you could ask, "What would flexible working arrangements look like if I was hired for this Laboratory Assistant position?" This second question is more likely to get a longer answer and affirms your interest in the role. We've given you some ideas to get started with in Activity 7.6.

Make sure that the questions you prepare are professional and contextually relevant. If you are applying for a laboratory position, it is unlikely that you can ask about arrangements for working from home. You also shouldn't be asking personal questions such as what their plans are for the weekend or how much they are paid. Revisit Activity 5.1 for some ideas of questions to ask if you are stuck.

> *"You don't even know what you're applying to half the time and then you like, they give you a description of the position, and you get into the position then... ask questions that kind of clarify what the daily routine would be in that position."*
>
> – Thomas, bachelor's degree focused on mathematics and PhD focused on fluid dynamics, start-up renewable energy forecasting.

Activity 7.6: Brainstorm Questions to Ask in an Interview

Activity goal: Create a list of questions that you can ask at the end of a job interview. Repeat for different roles.

Purpose and benefit: You will develop questions that will be useful to round out an interview. You will inevitably be asked if you have any questions (unless the interview is perhaps going over time), and you should have some ready to use. By having at least five questions you will be able to flexibly and naturally ask questions that are relevant to the role you are applying for.

Activity steps:

1. Open the job description for the role you are interviewing for. What are the role's responsibilities? Are there areas where you could ask for more clarity in who you will report to, or a specific technique?

2. Search in an internet browser for the company to refresh your memory on the company's products/services and competitors. What recent challenges or achievements has the company faced? Asking a question about these topics is also an opportunity to show off your commercial awareness (Chapter 3).

3. Look on the company's social media to see if the company has posted about any recent training or professional development programs. You can ask questions about further education to show your willingness to learn and interest in continuing to improve within the company.

4. Use social media and the company's website to investigate the company culture and management style. You could ask a question about mentoring programs and team hierarchies, which could help you decide whether the company is a good fit for you. Asking about teams also reinforces your skills in this area – but a question along this line would be inappropriate if you are applying for a role where you will work very autonomously.

5. Look into the career backgrounds of the people on your interviewing panel. You could ask them specific questions about their background – perhaps they have advice for a new graduate coming into the industry.

6. Make sure you have developed at least five questions that cover each of the above topics. Some questions will be quite universal and suitable for asking at multiple interviews (e.g., Can you tell me about what a typical workday looks like for someone in this role?) and others will be company specific (any questions about values and competitors for example).

Takeaway: By following these steps, you can develop a bank of questions to ask during an interview. Having these ready will help you to feel more confident in the interview, and have the bonus of forcing you to educate yourself about the company more carefully than when you wrote your three application documents.

After the Interview

At the end of the interview, you should have gotten a good feel for whether you will be considered for the position. Usually when interviewers ask about your availability, or mention that they are going to call your referees, this means that you have a good chance of getting the job. In the latter case, you should get in touch with your referees and let them know that they may be contacted. Inform your referees of the type of role you have interviewed for, and what skills you would like them to talk about to the hirer.

Follow up with a thank-you email to the interviewer to express your appreciation for the opportunity and to reiterate your interest in the position. This does not need to be long, just a quick note to remind them that you interviewed and that you are still interested. If you mis-stepped during the interview or forgot to share a key piece of relevant information, you can also mention it in your note.

Also, reflect on your performance during the interview (see Chapter 2 for structured reflective activities to help you). Take some time to think about how you did in the interview. What went well? What could you have improved on? This will help you prepare for future interviews. If there were some tricky technical questions that you weren't sure of the answers to, what can you do to upskill yourself in those areas?

So, You Didn't Hear Back

If you haven't heard from the interviewer after a week or two, send a follow-up email to check-in. This shows that you're still interested in the position and that you're eager to hear back. Even if you aren't going to be offered the position, they may put you on the shortlist to be contacted if other roles become available.

It is important to stay positive even if you don't get the job. Remember that there are many other opportunities out there. Keep applying for jobs and eventually, you'll find the right one for you. You may need to do a lot of interviews before you land a job, but that is ok, even if it is frustrating. Treat each interview as an opportunity to learn more about the company, industry and environment you are interested in – each one you do isn't a waste of time, even if you aren't successful. Following up with hirers if you haven't been offered the position can help you work out what your weaknesses are and where you need to improve. Were you one of their top candidates but someone else happened to be slightly better skilled for the position?

"I witnessed seeing my peers apply for many job applications and get rejected for just as many. I think having that process is important for building confidence for the interview process, but also to gain feedback as to what could help you next time when applying for a role."

– Sophie, biochemistry and molecular biology graduate and current bachelor's student in nutrition sciences.

Online Interviews

Many interviews now take place online *via* Zoom, MSTeams or Skype, rather than in person. Preparing for an online interview is similar to preparing for an in-person interview, but there are some additional steps you can take to ensure a successful virtual interview. Here are some tips to help you prepare for an online interview.

Technical Preparation

Ensure that your internet connection, webcam, and microphone are working properly. Be aware there may be a slight delay in communication due to internet connection or software issues. This can impact the flow of the conversation and may require more patience and attentiveness from both the interviewer and interviewee. Test with a friend or family member to ensure that you are comfortable using the technology.

If the platform being used to connect is different to your usual software, make sure that it is working and that you know how to use features such as using a virtual background, setting your pronouns and how to mute yourself. Have a backup plan in case of technical difficulties, such as a phone number to call if the video conferencing software fails.

Before the Interview: Set the Scene

Choose a quiet, well-lit space for the interview. Ensure that there is no background noise or distractions that could interfere with the interview. Make sure that your surroundings are clean and professional and remove any clutter or personal items that could be distracting. You should do this even if you intend to use a virtual background during your interview – you never know when technology will fail, and your dirty laundry will be exposed to your interviewers. If you wear glasses, check whether anything is reflected in them.

Dress professionally, as you would for an in-person interview. Avoid wearing bright colours or busy patterns that may be distracting on camera. Be conscious that in an online interview, body language can be harder to read and convey and make sure that your camera is set up to show your head and shoulders clearly and allow for space so that you can use your hands when you talk.

During the Interview: Focus and Connect

Make sure that your space will be free of distractions. Put your mobile phone on silent so that it does not disturb you, but keep it with you in case there are technical difficulties and you need to interview *via* the phone. If, like me, you have a lot of tabs and programs open on your computer, you will need to close them all. There is nothing like an unread email ping or FB alert popping up on your screen to distract you from the flow of the interview. Do not have notes. It is fine to write some notes beforehand to remind yourself of how to tackle a range of interview questions, but do not have them on your screen at the same time as the interview. You might be tempted to read from them, and that can cause your voice to go flat and expressionless.

Remember to make eye contact, smile, and speak clearly during the interview, as you would in an in-person interview. It may be more difficult to make eye contact or gauge the other person's reactions, which can affect the tone and dynamic of the interview.

After the Interview

You can wrap up the interview just as you normally would for an in-person interview. Follow the same process of sending a follow-up thanking the interviewers for their time, and reiterating your interest in the position.

Where to Next?

You have read about using the STARR technique to address key selection criteria and how to prepare for interviews. You'll need to revisit this chapter quite a few times each time you apply for a new role. As you gain experience, though, the approaches you have read about will become more automatic and you will likely find yourself doing them without thinking about the steps as consciously.

We can't emphasise enough that you need to prepare all of your application documents specifically for each role you apply for. Failing to do this will limit your chances of being offered an interview. As I suspect we all know, being unprepared for an interview can severely impact your confidence. This applies during the interview as it impacts your ability to answer confidently and lowers the chance that you will be able to address an unexpected or difficult question. But it also applies after the interview. While we all try to bounce back from a poor interview experience, it is difficult. The best path is to put lots of effort in beforehand to be very prepared. In this case, even if something goes wrong, you have peace of mind in knowing you literally tried everything in the book.

We both started our working careers not knowing how to prepare well for interviews, and it was not nice. This is one reason we wrote this book. We want to save as many students and graduates as possible the stress and drama that comes with not knowing what to do. Will this book make the process easy? No, not really. Interview preparation is still a lot of work, and there is no magic fix to the reality that no one can possibly be the best candidate for all the jobs advertised in their area of interest. So, prepare well, understand the market as best you can, but also be kind to yourself. Applying for jobs is genuinely hard work. From the both of us, best wishes on finding your spot in the job market.

Check Your Progress

In the first chapter you likely filled in a short survey to gauge your perception of your employability, which will have been emailed to you. This was called a self-perceived employability questionnaire (Activity 1.1). The subject on the email will have the title of the book in it. If you are ready to check your progress, use this QR code to recomplete the questionnaire and it will send the new results. Compare these results with any you have been sent previously by doing the following:

1. Bring the results up together and compare your results. If you are later in the book, you might have more than two. You can compare which results you want, or even all of them. Transferring the data to an Excel spreadsheet will help if you have many answers and are inclined to do that, or if you need an activity to help develop your Excel skills.
2. What are the differences? Sometimes you will feel a bit different on a different day, so not every value that changes by one point is a significant change. Maybe you had a good day or a bad day. Can you see any trends or big differences? Do any of the changes feel particularly true? Even if it is just one point.
3. Often when people learn more about a subject, they first loose a bit of confidence as they are starting to learn about how much there is to learn. Has that happened to you? This is an important process to be

Continued

> **Check Your Progress – (*Continued*)**
>
> aware of. Maybe you aren't feeling as comfortable with your employability but that's because you had to realise how much there was to know in order to be able to go out and learn it. Well done, that was an important realisation.
>
> 4. Now that you have stood back and thought about your overall perceptions of your employability, are there any plans you think you might make or change? Or maybe something you have realised you want to prioritise or deprioritise? If so, make a note of these and set a calendar reminder once a week for the next month to ensure it doesn't get forgotten.
> 5. Enjoy the next chapter!

References

1. A. Heine, The Best Time to Interview for a Job (Factors to Consider), *Indeed*, July 13, 2024, https://www.indeed.com/career-advice/interviewing/best-time-to-interview.

Looking for a Job (Before and After Graduation)

DIMANTHA HARSHAPRIYA[a] AND
REBECCA YEE[b]

[a]Norwegian University of Science and Technology,
Norway; [b]GPA Engineering, Australia

Introduction

In this chapter, we will explore the process of *finding* a job. How can you search for a specific role or even become more aware of what jobs are out there? Do you even know *where* to look? This chapter will focus on the search itself. The process of applying for a selected job (through resume writing and interviewing) are addressed in Chapter 6: Resumes and Cover Letters and Chapter 7: Addressing Key Selection Criteria and Interviews.

These search tactics are most relevant to finding an early career role in STEM fields. However, there are many tips that can still apply to any type of job or even different stages in your career. Entering the professional world for the first time can be challenging. It's a winding journey, just like the path in Figure 8.1. This chapter will help you develop the tools to start this journey. We will discuss where you can find suitable jobs, as well as what to consider in your search, and how to make the process easier for yourself.

If you haven't already, check through Chapter 4: Networking, as this is usually the best way to start looking even if you don't yet know what you are looking for! Even if you are still studying, it is never too early to start looking to see what type of work may be available. You may continue this search all throughout your studies and even for a period after graduation.

Be alert to opportunities that your university or school may provide. There may be networking events or internships that can introduce you to various fields or professional environments. Work placement programs could give you valuable insights into *different* types of jobs before you even graduate. There are numerous paths to starting your career. While your first job is not an 'ultimate destination', preparing yourself with these tips and strategies will help you approach job hunting with confidence.

After years of classes and progressive coursework, many students may hope that entering the professional world would be as automatic as going up to

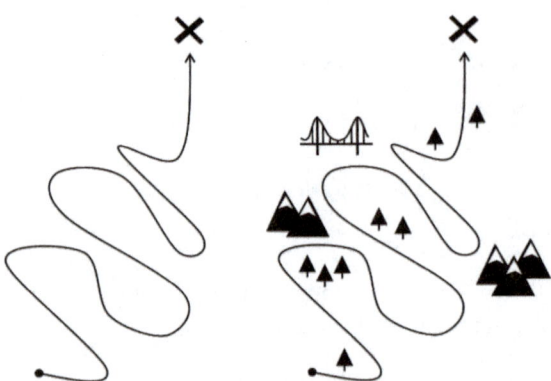

Figure 8.1   Your career path may take a winding route. Knowing what to look for and where will help you navigate your journey.

the next grade. Some students don't even think about work until after they have finished their studies. Some may expect a job offer to land at their feet. On the other hand, many students play the 'numbers game' and submit hundreds of resumes without getting a single interview. With the preparation and guidance from this chapter, you can hopefully avoid both these cases by focusing your efforts on what you want and reducing your overall stress.

Sometimes it may feel like you are under pressure to decide on your career immediately after school. However, it's important to remember that a career is a journey, not a destination. Finding a path that you're happy with may take some time and self-exploration. You may end up trying many different roles and types of employment, from casual, full-time, or contract work, for example. Together, let's take a closer look at how we can unravel and understand the process of looking for a job.

What Type of Job Do I Want?

Thinking about your skills and what you want to do is often the first step before you even start looking. Chapter 2: Transferrable Skills and Reflection will be useful to do in parallel while you explore the job market. What types of skills are needed for the jobs that you want? Exploring the job market broadly can also help you identify what kind of jobs appeal to you. It is then very important to select the roles you really want so you do not waste your time, energy, and mental sanity in applying.

Hopefully, if you are reading this guidebook, you have already developed some professional interests. Perhaps you can list a few technical skills, but what job title relates to that type of work? What kind of company would employ you? This section will help you start understanding your own career motivations. Which of the people in Figure 8.2 might you be?

What kind of jobs would even suit you? You might not even be aware of the jobs you can do with your degree!

There are many aspects to a job, some key ones are laid out in Figure 8.3. In most cases, it means that you exchange your time and labour for money.

Figure 8.2 Different people will prioritise different goals.

Figure 8.3 What is important to you about your work and career?

But there are many factors that will impact your **job satisfaction**. I will pause here to note I'm not talking about a "dream job", which may just be that – a dream! Especially when first starting out, with little experience or practical insights, you may not be able to get everything you want from your first job. However, this is a useful place to start thinking – what is **most** important to you? There are aspects to work that are more or less important to some than others. What are your priorities when looking for a job?

Keep in mind that what may be important to you now (*e.g.,* huge amounts of money) may change in the future. This is completely valid. Everyone has their own preferences and priorities can change a lot throughout your life. Your career journey is not set in stone, and flexibility is a key aspect to embrace. You may even recognise that your own mind is likely to change as you grow and mature!

> *"I've worked at a couple of places and in the move to my current company I was looking for a low-stress job without too much overtime and decent pay."*
>
> – Nimaz, software engineer with a UK education, working for a company in Sri Lanka.

STEM careers can be very rewarding and there are many opportunities to choose from. STEM skills are also highly transferrable, and graduates can be employed in almost every industry sector. So how can you narrow it down? What else should you consider when looking for a job? A simple way to start when there seem to be endless choices is to eliminate the options that you definitely **do not want**. It may be useful to think about what you would never want to do. For example, if you are a vegan for ethical reasons, it's unlikely that you will want to work at a pharmaceutical company that does animal testing. Perhaps you have excellent analytical skills and enjoy working with data, but you know you would rather work for a community organisation rather than a military project. Activity 8.1 will guide you through understanding some of the considerations that may impact your career path.

Some job markets and STEM fields can be very competitive. You don't want to waste your time or energy applying for everything under the sun. The more you recognise your goal(s) and priorities, the easier it will be to find a good fit. The following sections discuss a few issues that you may want to consider about your own job satisfaction. Use them like this map in Figure 8.4 to find your way and be aware of major features on the road ahead.

Activity 8.1: Consider Your Priorities

Activity goal: To reflect on your own motivations and identify the values of your work ethic.

Purpose and benefit: Understand and prioritise the work-related values and personal motivations to make better career choices.

Activity steps:

1. Look through these priorities and note the most important factors for you (1 – most important, 5 – least important):
   - opportunities for career progression
   - training and professional development provided
   - travel as part of your job
   - need to stay in your current location/area
   - need to stay in your current country
   - flexible hours
   - working from home options
   - high financial remuneration
   - intellectually challenging work
   - friendly and sociable work environment.

2. There are more considerations than the options listed in part 1. What other variables are important to you? List three of these by reflecting on how you want to work. Doing some internet searches might help to investigate what others see as important variables for their work.

3. Write out and score your three new variables according to the scoring in step 1. Having looked at how you have scored the 10 factors in step 1 and your 3 variables in step 2, did thinking about them change any considerations? Or maybe the process just solidified views you already had?

4. Now try writing a list of things you would never accept if there are additional points. Did that help or were they already accounted for in steps 2 and 3?

5. Over time as you think through other aspects of what you want out of a job, and possibly gain work experiences, your priorities will change. Set a calendar reminder to revisit this activity and revise your thoughts, including if you have other variables that are now important that weren't important earlier. If you are very actively working on career development, try to do this every couple of months. If not, twice a year is probably a good frequency. Note down the Activity number in your calendar to save time.

Takeaway: Everyone has aspects they want or don't want from their career, but many of these things are unknown to us when we are students or recent graduates. By being very conscious to reflect on what you want and why, what works for you in a workplace experience and what doesn't, you will more quickly get to an answer and insights related to what jobs will better suit you.

Figure 8.4   Being prepared makes a world of difference. Photo from Tabea Schimpf on Unsplash.

> *"I was a microbiologist doing testing. I was there for probably seven to eight months. To be very honest, I didn't enjoy it, I did not enjoy the job. And I think this is a personality thing. It's mostly because I'm quite extroverted. And for me, to stay in the lab where you are constantly pipetting like samples, hundreds and hundreds of samples every single day, but it doesn't work really well for me at the same time, the fact that the company culture wasn't really great didn't help as well."*
>
> – Joelene, Bachelor of Biomedical Science, now working in marketing.

Money and Benefits

It is common for many graduates to focus on the pay when looking for a job. Students who have experience with casual jobs may be familiar with being paid per hour. Of course, at first glance, it makes sense to want the $60 per hour job over $30 per hour! But **wait**, what if that $60 per hour job means you have to travel much further, and pay for your own car and fuel, and maybe have a contractor agreement that means you need to arrange your taxes or retirement? Maybe the higher paid position is only temporary, with a requirement for the job to be finished within a few months. Perhaps the $30 per hour job includes a substantial end-of-year bonus based on the company's overall profit. Maybe the lower pay is intended for several months of a probationary ('trial') period and then switches automatically to a more standard salaried role.

Full-time or professional roles are often provided with a salary. Salaried employees usually earn a specific amount over an entire year, as opposed to earning by the hour or for each piece of work. A salary is typically paid in fixed increments throughout a year and includes an agreed amount of holiday and sick day benefits. These ***entitlements*** can provide secure cash flow, where you will still be paid even if you take time off due to illness. However, you may also be expected to work unpaid overtime to get the job done.

In addition to financial compensation, many companies offer other benefits. Employee benefits can include health insurance, retirement plans, paid time off for sick leave or parental leave, *etc*. Such details are often overlooked by new workers that do not realise that leave entitlements can accrue (bank up) the longer you work for that company. You can find more information about employee benefits and policies and legislations in Chapter 3: Commercial Awareness. Many companies have their employment benefits listed on their public website. The recent global pandemic was a good example of how different companies and company policies impacted their employees.

As a new graduate, it may be difficult to recognise what is 'fair' or 'market rate'. Keep in mind that what is 'standard' can also vary between countries and industries. There are several websites that can give an indication of average rates for various roles and job descriptions. Remember that financial benefits are only one thing to consider. Keep in mind that you are also likely to increase your salary as you progress and develop your professional skills. Can you relate to any of the people in Figure 8.5?

Figure 8.5 Discussing wages and salaries can be viewed differently around the world.

Company Culture

> *"It's not just about you and your work, it's you as a person as well and whether you'd be the right culture fit."*
>
> – Lane, chemical engineer, previously worked as an environmental engineer in renewable energy sales, then an efficiency engineer and finally a solar engineer/designer.

Employers and managers often talk about "finding the right fit" but remember that we are still social beings and the people around us, and how we interact with them, is meaningful. Company culture describes how the employees behave, interact, communicate, and work within a particular company. What does effective teamwork look like for you? Do you have a preferred style of communicating? What gives you a sense of value?

Company culture can have a huge impact on the quality of your work life, relationships with peers, mental well-being, work-life balance, and overall career progression. Maybe you don't even want to be part of an ambitious culture at all and simply want to balance your time with other commitments. Do you want to feel secure in your role, having the confidence that the company will not fold within a year? Do you want to feel challenged and recognised for the work you do? These feelings are difficult to pin down and can vary significantly between different groups, organisations, or industries, depending on many factors.

Work-life balance is another quality that is highly subjective. For some companies, there is a culture that working long hours is fine if the pay is rewarding. For others, the company culture supports more flexibility with the trade-off that their staff accepts lower pay. Other people prefer to work for a company that fosters social connections, such as paid conferences, formal networking events, or Friday lunches. Everyone has different views on balance and understanding your own priorities will help you find an organisation that will align and support your well-being. Use Activity 8.2 to practice how to assess the reputation of a company.

Think about what matters to you. Does the role expect you to work overtime? Do you have to work on weekends? Will you be asked to attend team-building or social events organised by your manager? It is common for most graduates to focus on finding any job as soon as they can, but those that take the time to dig a little deeper first can find something that works better for them over the long run.

Activity 8.2: Assess a Company's Reputation

Activity goal: To identify and explore aspects of an organisation's reputation.

Purpose and benefit: Develop an awareness of commercial entities and how an organisation's reputation can influence your career or career decisions.

Activity steps:

1. Visit the website of a company that is related to your field of study. This could be an equipment supplier (look at any brand tags), a service provider, big or small. An internet search will help you find a few to pick from. Bigger companies tend to have more of this information, but some small companies are also great at making sure the benefits of working with them are clear to see.

2. Does the Company have a "Mission Statement" or public "Values"? Take a copy for reference if you can find something that says what the company sees as important.

3. Does their Careers page describe any training programs that would help support your career growth? What might be useful?

4. How does the company consider mental health and work-life balance? Do they have a page to explain benefits to applicants?

5. Search a news website for the company's name. Has there been any media articles about the company's financial, legal, or environmental impact? Remember, this isn't just an ethical consideration, companies in financial, legal or environmental trouble can't always keep their employees. Make sure you are writing notes as you go so you can refer back to which company does what. These will blur together very quickly.

6. Does the company's website align with their media reputation? If it doesn't, this is a big warning sign. It is easy to put together a nice page about careers but much harder to actually be a good company to work for.

7. Do you know or have any connections to someone that works for that company? Can you contact them to ask about their experience with the company culture? While it is not frequent to know someone who work(ed/s) at a particular company, it is a good activity to make sure you ask around to see what people think of companies. If a company treats its people badly, it is unlikely to be a secret. It doesn't have to be a close acquaintance; a neighbour might have a friend that had a bad experience, for example, or a second cousin, or an aunt's friend, *etc.*

Continued

Activity 8.2: Assess a Company's Reputation
– (*Continued*)

8. Repeat for at least two more companies. Looking at a range of companies is important so you can compare and know the standards in the environment you are looking at.

9. This will add up to quite a bit of work, so set a reminder to come back if needed. Also, repeat when you are thinking about a different work environment, or after 6 months, as companies and the employment environment change.

Takeaways: It takes time to map out good and bad employers and learn from people about their experiences. In different regions, there will be different approaches to whether people will talk about poor experiences with a former employer. However, a universal trend is that stories do get out about the best and worse employers. In many places, people will spend 15 minutes happily making sure you understand how much they love or dislike an employer. With opinions maybe being more honest after someone leaves the company. Take advantage of this and use others' insights with your own research to build a picture of your work or future work environments.

Recognising a 'good' company culture involves paying attention, but experience also helps. You will build this by observing how employees interact with each other and with management. Positive signs include open and respectful communication, collaboration, and mutual support. No one is being left out or put down. Is the company committed to employee development and work-life balance? Do they have initiatives like training programs, mentorship opportunities, and flexible work arrangements? Additionally, you can look for a clear mission and values that align with your own, as this indicates a shared purpose. You can also seek feedback from current and former employees. Leveraging your professional network, including LinkedIn connections, allows you to connect with individuals who have worked or are currently working at the company, enabling you to ask pertinent questions about their experiences. Furthermore, exploring the company's social media presence can offer glimpses into their values and how they engage with their audience. Assessing these factors helps identify a company with a positive and supportive work environment, fostering employee growth and well-being.

The world has become a lot smaller now, and we often hear people use the term "global village", especially regarding STEM careers where qualifications are often internationally recognised. Globalisation has significantly changed the way people work around the world. Many workplaces have become more diverse, with people from different backgrounds and ages coming together. Some traditional fields that were previously dominated by one gender (*e.g.*, engineering or teaching) are changing rapidly with this generation. This means that working with colleagues from a wide range of backgrounds and identities is becoming more common. Working for a company that recognises and respects diversity – such as nationalities, ethnicities, religions, and languages – may be extremely important for your work-life balance and overall mental health. It is worthwhile to see if you can tell whether the workplace has created a good working environment for all its employees, regardless of their cultural differences.

> *Hint:* Ask your friends and family what they enjoy about their job. What do they dislike? Hearing stories about employment experiences will help you identify red flags and signs of a healthy work environment.

Career Development

In addition to financial compensation, your professional development can also be enhanced by the right employer. Growth opportunities could include in-house training offered by your company, or the opportunity to travel for work or even be paid to attend conferences. Some companies have a specific pathway that gives you a clear step-by-step outline of job roles. Others may give you the chance to shape or choose your own work. If you are eager to advance your professional experience, it is important to check if such opportunities are offered in your workplace. For instance, did any of the senior staff start off working as a technician or low-level operator? This could show that you can 'climb the ladder' within that organisation, or that your company values loyalty and will promote staff internally. In other cases, all the senior staff may have PhDs or be chartered professionals, showing other company preferences.

Different companies have different systems for promotion and career development. Some places may have a formal mentoring program, while in others your boss looks for those doing great work and recommends you for promotion. In some roles, you will have to move or transfer to another department if your starting role is too defined or limited. You may be expected to move towns to get promoted as the senior staff are expected to be located in the city of the head office.

When considering your professional development, also think about industry bodies that accredit or 'charter' certain roles, like accountants or engineers.

It is never too early to get involved, and many professional organisations have discounted student memberships. These groups may host events or provide additional industry information to expand your commercial awareness (Chapter 3), which will help you explore more career opportunities.

> *"If you are trying to make a decision about your career, I'm a big fan of mentors, different mentors so you can see those different perspectives through their eyes. Talk to someone who has worked in that environment before you make decisions, so you know what to expect."*
>
> – Alix, political science graduate with a law master's and two decades of experience in various areas of government and not-for-profit organisations, specialising in policy.

The reputation of your employer can also significantly influence your future career path. For example, if you work for a company that gets sued for significant environmental damages, this may impact your chances of getting a future job in ecological conservation. Reputation is difficult to gauge as a student or graduate.

Try to do a background check on any company before you accept a job, or even before you attend an interview. Have they been in the news recently? Have you heard of the name of the employer before? Are you familiar with some previous projects done by that organisation? What are their plans in terms of expansion? Does the company or the institution actively take part in projects that help the world? Does the organisation collaborate or support furthering your education or qualifications? All these questions, to different degrees, can provide insight into how a company can help develop your own career.

So, Where Can I Find a Job?

Throughout history, most people got their first job (and lifelong career) through apprenticeships, where they would follow an experienced person and learn on-the-job to be the town's next baker, butcher, or candlestick maker. Your grandparents, or even your parents, probably found their first job in ways that would be impossible or even laughable today. It is most likely that they had a family friend or contact that got them in the door. Has anyone told you to take a printed copy of your resume and knock on company doors until you find a manager to hand it to? Perhaps you could do this for a casual retail or hospitality job but, especially for a qualified STEM role, this approach will not get you far in professional settings. Explore how other people got their first job with Activity 8.3.

"The biggest hurdle was definitely just how do you find a job? Where do you start?"

– Alex, arts/science graduate with a biochemistry major and master's in bioethics, now a lecturer in indigenous health.

Today, the world has become smaller, and the job market has become global. Expectations for a well-formatted resume submitted online have become the standard. Recruitment, especially for larger organisations, has become heavily automated. This can feel daunting, but it just means that it is even more important for you to take your time and only apply for the jobs most relevant to your qualifications, skills, and interests.

In this section, we will guide you to the start of your job search by using a few common sources. At this stage, don't worry about the application process. It is important to take some time to see what is out there.

"Finding the right job, not just any job."

Why Is It Called Job Hunting?

When a lioness is hunting, she does not charge in willy-nilly snapping at every buffalo. She first watches, observing the options, and then selects which prey to chase. She doesn't want to waste her energy and nor should you. A job hunt follows a similar process. By taking time to observe and think through your choices, you will be better able to select the jobs that are best worth your effort to chase.

"My college and professors didn't have a lot of suggestions of where to apply to work, so I went to a staffing company, which worked well. So did a lot of my friends. Therefore, when I moved to Europe I used a staffing company again. I highly recommend it for people who don't know where to start, if that sort of service exists in their environment."

– Alex, research scientist at a microbial protein production firm, born in the US, has lived, studied, and worked around the world.

Activity 8.3: How Did Other People Get Their First Job?

Activity goal: To learn about early career job pathways. This task will take about a week.

Purpose and benefit: Learn from the experience of others. Find opportunities to apply for a job.

Activity steps:

1. Think of people who have jobs that seem interesting to you. It may not be something you want to do yourself but perhaps they seem to enjoy their work and would be considered to have a successful career. The idea is that you might learn something from their journey not their exact destination.

2. List several people that you can contact – either speaking to them in person or by email or phone.

3. Reach out and let them know you are working on career planning. Enquire whether they would be happy to answer a few questions (phone, email or in person). It might help to say you are working through a careers book and this is an activity in the book.

4. If they are happy to talk (most people will be, the biggest challenge is usually convincing people they can help), ask them how they got their first job? How did they hear about it? How did they prepare? *etc.*

5. Reflect on these methods and think if they would work for your situation. Which steps are relevant to you? Remember, you may have to modernise to update the approach. Someone who mailed a lot of paper resumes out to well-researched companies or individuals likely wouldn't do that now. The useful advice might be how they worked out who to target and do they have any advice. Did they build relationships before sending? Do they have advice on their approach with hindsight? *etc.*

6. Make a mental note to keep an eye out for new people to talk to. These might be in your existing network, but this is also a great networking topic for conversation when you are at an event, casual, formal or impromptu.

7. Schedule some time to repeat steps 1–5 in 3–6 months or even sooner if you want to start applying for jobs very soon.

Takeaways: People are often very happy to share their journey through their career. It's important to talk to a range of people and get a range of impressions about what people do. Also, it is important to keep in mind all advice is not equal. It is a common human trait to either over-value or under-value our experience, depending on personality. Selecting who you want to talk to based on your respect for their career journey is valuable. But also taking additional opportunities that come up along the way can also help. Just remember to keep your critical thinking cap on to identify what advice is for you. It isn't easy, but it can help you understand career development and, very importantly, you will soon see there is never just one way.

Online Job Search Engines

"I think there's plenty out there. You just got to find it. Be willing to sort of step away from the very typical career options and have a look for something adjacent. Yeah, there's plenty out there. It's just about figuring out what that is."

– Nigel, science/entrepreneurship majoring in physiology, working in marketing for an animal aid organisation.

Online search tools, like that seen in Figure 8.6, are almost second nature to most of us these days. Never before has the job market been so easy to access through online employment websites. These platforms display many different job listings from different employers or recruiters to list their open positions to the public. For example, LinkedIn is a professional networking website that also has a Careers page, which can help connect your online presence with job opportunities. A good first step is being aware of what job search engines are available to you.

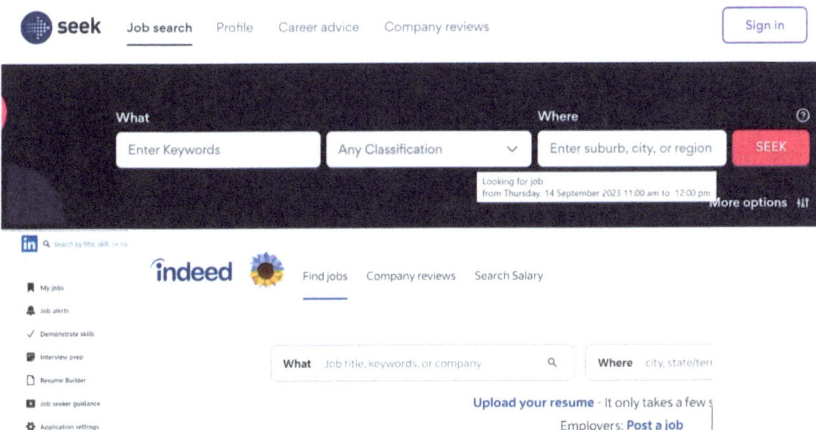

Figure 8.6 Online job searching can take place at a variety of websites including Indeed, Seek and even social media platforms such as LinkedIn.

> **Box 8.1   Job Scams**
>
> As with all online use, you should be aware of potential scams. Fake job descriptions can be used to target scam victims. This is where your STEM critical thinking and analysis skills become important. How do you know that job description is reliable? Is the company real? Does the job sound too good to be true – with high wages, flexible hours, and no experience or qualifications required?
>
> Remember, you should **never** be asked to pay a recruiter or HR representative. Be cautious if anyone asks you for money during the hiring process or 'sends' you money so you can perform a task as a 'test'. See Chapters 6 and 7 to get a better understanding of the standard recruiting processes (*i.e.*, what happens after you apply). This will help you be aware of any red flags that a reputable company would not do. Some platforms, such as Work180, verify every company they post. This helps to ensure that the jobs being advertised are genuine.
>
> To verify the authenticity of a job description found online, you can check the employer's official website and online presence for consistency and credibility. Also, you can cross-reference the job posting with reputable sources, such as Indeed, CareerBuilder, or the employer's official website, to confirm its existence and compare details. You can also contact the company or recruiter directly to enquire about the position and ask for more information.

How to Use an Online Job Search Platform

Many students and graduates just type in their degree name when first searching for a job. However, the professional world does not align jobs with such qualifications (see Chapter 3 for further discussion on how industrial sectors are shaped). Someone who majors in Chemistry will struggle to find a suitable role if they only search for "Chemist". Someone who wants a job working with numbers and data is unlikely to be successful searching for "Mathematician".

Just like searching for relevant and appropriate scientific papers, selecting the right search terms and keywords that match your preferred job is a skill in itself. There are often filters you can use that specify factors such as location or pay range. Activity 8.4 is useful to start learning how to narrow the search for jobs that are applicable to you. You may want to repeat these searches regularly, especially towards your graduation date, as new jobs will be posted all the time.

You can learn a lot from looking at other job descriptions. For example, if you put in microbiology as your search term because that is your degree,

Activity 8.4: How Do You Get the Most Out of Online Job Searching?

Activity goal: Learn the relevance of keywords and search filters.

Purpose and benefit: Broaden your awareness of types of jobs relevant to your skills and interests.

Activity steps:

1. Pick any job search board. Type in the name of your degree and/or majors and search for jobs.

2. Assess the relevancy of the search results based on your interests and skills, but also your future qualifications if you are a student. You can consider factors such as transferable skills wanted, required qualifications/experience, and the alignment of the roles with your career goals (location, training, benefits, type of company, *etc.*).

3. Select three jobs and open the job descriptions in full. Be sure to check if there are additional documents linked in the advertisement. It is common that the advertisement is not the full job description. Note the advertiser and look them up. This will help you understand who is posting the roles and therefore at least one person likely to be reading your application and on the interview panel. If the contact is HR, is there another name or title in there? For example, reports to the "deputy head of fieldwork".

4. Perform an internet search with this title and the company name and see who you can find. Feel free to view them on LinkedIn. It's normal that someone looking for a job at a company looks at who is hiring. They like to see that there is interest.

5. Highlight related terms listed in the job description to further investigate. Copy and paste these into a new search tab in the same job search engine and review the results. What did you find? Was there anything of interest? Did you discover any new titles that you can use in future searches? Be sure to be noting down any points of interest, as these are easy to forget.

6. Now repeat steps 1–5 starting with more precise terms. This will usually be a common job title that your environment uses. Now that you have worked out a few common titles that fit jobs you might be interested in, what do you find? If you aren't finding jobs, you haven't succeeded in getting the search terms right yet. This is a common issue. Try a few more cycles of steps 1–5 and talk to friends or colleagues if you still don't have success. And remember your institution likely has someone to help too.

Continued

Activity 8.4: How Do You Get the Most Out of Online Job Searching? – (*Continued*)

7. Repeat these steps every few months, even if you are not actively looking for work. This is a useful activity to stay on top of industry changes and build your knowledge of search terms, trends and even the most common pay rates and conditions. This is all extremely valuable information to have.

8. Repeat any time you move your interest to different types of roles, or at least each year if your interests remain the same. More often if you are actively applying.

Takeaways: When you first start searching, it is really common to have trouble finding jobs that are suitable. I've seen many heart-broken students that are convinced there is nothing for them. But in reality, a few of the above tricks usually help them find relevant roles. Some will be of the wrong seniority, but these can still tell you about your field. Next month there might be only entry-level roles and no senior roles, it always varies so you need to keep looking. It's also good to see what a future role might look like. Treat this activity as a general learning experience about your possible future industry. Does the senior role look great or terrible? If you wouldn't be happy doing the more senior role, that's one more bit of information in understanding where you want to be.

what else is listed in the description as suitable degrees or qualifications? Who else can do that job and therefore what else might you do? Some roles won't be similar enough to give any insights (*e.g.,* someone could hire a chemical engineer for a chemistry job, but a highly technical chemical engineering job is unlikely to go to a chemist). However, you can start to build a picture of the language being used and what roles you might be eligible for in your environment.

Let's think about the microbiology example again. The advertisement might want someone with laboratory technician experience. They might not care that much about your major being microbiology. They might want you to have proven that you can function in a laboratory more. That tells you that if you want to get a job as a lab technician, it doesn't necessarily have to match your degree or major. Probably you can apply for general biology, biochemistry and even chemistry laboratory technician roles. Understanding this will help you work out where your opportunities are. Also remember it is usually seen as OK to call or email and ask if your degree/major/experience means that you suit the job if the advertisement is unclear (they often say "and related qualifica-tions", which could not be more vague). Whether you have done online job searchers or not before, start with Activity 8.4 to see what you might learn to help you.

> *"Most chem students and grads I've known have little to no idea what jobs are available to them, what the pay is like, which companies employ chem-ists, etc. They just know they love chemistry and want to keep doing more chemistry – and often hitch their hopes on being like the only professional chemists they know: their professors."*

> – Aaron, biochemistry and molecular biology graduate with research experience, not-for-profit experience, and experience managing grad-uate education programs at universities.

University Careers Centres

Pay attention to the walls and boards around your campus. Physical job advertisements on paper still exist, especially for casual or one-off opportunities, like washing dishes for a laboratory. Maybe the local school library needs someone to shelve books or run a STEM for Kids Club program.

Your school or university may also have an online job website that is only for students (see Box 8.2). These may include jobs on campus or research positions. Many institutions have a dedicated careers advisory depart-ment that provides a highly valuable (and free) resource. Importantly, these help resources are ***often available after graduation***. If you haven't heard of help available through your institution, don't assume that means there isn't help. It's not that uncommon for STEM students not to know about help available to them in the careers space. In some cases, these teams might also be quite small. In other cases, the careers team and STEM students seem not to end up meeting for a variety of reasons. Call your university central telephone line or email the general student sup-port line and ask for the careers help centre. You might end up having to explain a little more or ask in a few different places but persist and there is a 95% chance your institution has some help. It also might not be uni-versity-wide, it might be faculty or school based. In that case, approach-ing the school or faculty office is the path forward. Be aware that it is not uncommon for academics not to know what resources are available for students, so certainly ask a trusted academic, but if they say they haven't heard of a centre or office, don't assume that means that one doesn't exist.

Box 8.2   Career Centres

At your university, Career Centres are managed by qualified professionals along with student career ambassadors. These centres are dedicated to enhancing the employability prospects of students and graduates, equipping them for the challenges of the professional world. Within these centres, you can find valuable resources, such as online materials, resume templates, and career guidance. You can easily connect with advisors through your student portal or various social media platforms, or arrange appointments to discuss your career goals.

In addition to these resources, Career Centres and student groups frequently organise industry-related events. These events encompass expert panels, Q&A sessions, and networking opportunities, offering students the chance to meet and converse with industry professionals about their career aspirations.

The University's Career Centres host a variety of workshops and seminars, both in-person and online, catering to the needs of both students and graduates. These sessions cover a wide array of career development topics, including personal branding, networking, resume and cover letter crafting, as well as interview skills. These centres often facilitate work experience programs, internships, mentorship opportunities, and even volunteering initiatives, all designed to enrich your professional journey.

– Liz Hendry, Careers Centre Manager, Charles Darwin University.

University sourced roles could give you exposure to bigger companies that do work in your field of interest. Maybe you meet someone that offers you another opportunity to build new skills. The experience itself could add to your resume, perhaps including field work that you hadn't done before, or give you insight into how some organisations are run. It is simply a matter of taking some time to stop and have a look at what is out there. There are many guides (like this book) that can help.

Networking When Applying for Work

The benefits of networking have been discussed much more in Chapter 4. A skill in itself, networking also provides an informal way of learning about the job market. Keep in mind that networking should be about making genuine connections, with no end goal, purpose, or expectations beyond that. However, there is no doubt that many, if not most, jobs are found through networks and 'who you know'. You never know who you will work with later on. For example, the editors of this book were tutors together years ago.

Networking also includes socially connecting with other students at university or even friends from sports clubs. While there is little chance someone will offer you a job right on the spot, it is possible that they can make you aware of an opening or even just introduce you to a company that you would be interested in. Even just learning about other people's jobs and experiences will give you food for thought in your search.

Internship Programs

You might have heard the word "internship" being tossed around. These differ from casual or part-time jobs as they are specifically a short-term (*e.g.*, a few months) program that is aligned with your field of study. An internship is a structured working opportunity, often coordinated by the university, and typically open to students during their studies. These programs can also serve as a formal part of education or training, sometimes counting for credits instead of a 'normal' unit. You may be asked to complete a report or assignment specific to that internship. Many internships also assign both an academic supervisor as well as an industry supervisor, to ensure you are supported in a new work environment.

Internships can be very useful for both job seekers and employers. It typically spans a short period, so there is less commitment if you do not really know what you want to do. The aim is to provide entry-level experience in a specific job role, field, or industry relevant to your studies. During an internship, both the employer and the intern understand that it is a learning experience, and the intern may not be compensated at market rates and may require more training. This understanding fosters an environment where interns are encouraged to ask questions, seek guidance, and explore different aspects of the job or industry. Internships offer a unique chance to observe and learn from experienced professionals, expanding your professional network and gaining mentorship along the way.

Internships can sometimes lead to full-time jobs or positions if the company values the results of your work. From an employer's perspective, hiring the 'right' person through interviews is tricky. It's difficult to tell if someone is who they say they are in a short 30-minute conversation. How do you know they will fit in with the team? Can they really use the skills they claim to have in their resume? Are you reliable? An internship helps an employer see how you work and communicate on a regular basis. Even if the program does not lead to a more long-term position, the experience undoubtedly will stand out on your resume and give you more to discuss in your next job interview. Most employers will favour a graduate with some work experience.

Finding Internships

Internships are usually arranged through your university; some degrees may even have this as a formal requirement to graduate. It is essential to search actively for internship opportunities on your own as they may not land in your lap or be pushed forward like with an internal campus unit. Start by asking your lecturers or friends if they know of any placement programs. Directly check the websites of organisations that work in your field. Internships, or vacation programs, are less likely to be advertised on major job boards.

Get in early! These programs typically run seasonally over the summer holidays or in line with semester dates. However, applications and the selection process may be 6 months prior to the start of the program and if you want to develop or find your own internship (which can be very rewarding) this can take even longer. Internships can be full-time (Monday to Friday) or part-time (usually 2–4 few days a week). Use Activity 8.5 to help you work through what sort of internship you could do.

Internships can be very valuable in shaping your early career. They allow you to explore your interests (or find out what you **do not** want to do), discover your strengths, and identify areas for further development. The hands-on experience gained during an internship provides insight into how organisations work, what specific job roles do and how they interact, and will give you a taste of the overall working environment. Treat your internship as an opportunity to showcase your abilities, build relationships, and make a positive impression. This exposure can help you make more informed decisions about your next career steps and enhance your marketability to potential employers.

Internship: An internship is a structured working opportunity that typically spans a short period, often a few months. Its primary purpose is to provide practical, hands-on experience in a specific job role, field, or industry. Internships are usually offered to students and fresh graduates and can be either paid or unpaid. The emphasis is on gaining industry-specific skills, networking, and understanding the day-to-day responsibilities of a particular job role.

Field Training: A period of training that allows students to apply the theoretical knowledge they have acquired in an academic program within a real working environment. It is usually integrated into certain academic programs, particularly in fields like engineering or environmental science where students can work at a plant or out in the field. This provides students with exposure to real-world working conditions, practical applications of their theoretical knowledge, and an understanding of industry practices.

Activity 8.5: Finding Leads for Possible Internships

Activity goal: Find potential internships that would benefit your career development. This task will take 90 minutes.

Purpose and benefit: Have a better idea of where you can apply and what skills you might develop.

Activity steps:

1. First, brainstorm about where you might like to do an internship and make a list of three things you would like to get out of an internship.

2. Now think about what connections you might use to get an introduction. What charities do you know of that could use the help? Have you seen friends, classmates or family intern anywhere else? Make a list of three places/people you can approach.

3. Ask your Careers Help Centre or your local friendly academic if there are any formal internship programs run by the school, faculty, or university. Information on structure subjects will also be in the university handbook or an equivalent document that contains a description of all the units. Universities often have existing industry partners, but places are competitive.

4. Hop on the internet and search for programs in your area. There might not be a lot, but you only need one host. List any other research groups, companies, or institutions that may offer internships in your desired industry, OR that are offering internships that will likely develop skills you need (e.g., data analysis or organisational skills can be gained in any number of industries).

5. Note the dates any formal applications are due so that you don't miss out due to forgetting. These dates are usually well in advance, so looking the year before is an advantage.

6. Put a reminder alert in your personal calendar when applications are open.

7. Revisit Chapters 6 and 7 to get your draft CV prepared, refresh your LinkedIn profile (or build one if you haven't yet), and put in reminders so you can keep progressing on identifying what you want, where you can do a placement and when your paperwork is due.

Takeaway: Finding and applying for internships is excellent practice for applying for jobs. So, you can view the above preparation as helping you get an internship but also helping you get a job in the long run because of the practice with applications, potential interviews, having your LinkedIn ready, *etc.*

Volunteering

> *"I think it's hard to prepare for your first job, but I think the only way to prepare properly for professional experience is to gain professional experience. And, you know, it doesn't necessarily have to be your first job, but it could be something as simple as volunteering in those type of areas to kind of understand what it feels like."*
>
> – Marina, psychology and literature graduate working in HR (an increasingly common "irregular" pathway for psychology graduates).

Volunteering has been shown to have many benefits (Figure 8.7) – giving you many opportunities and skills that do not have to be aligned with your profession. In fact, you may continue volunteering on weekends while having a career in a totally unrelated field. Volunteering tends to be easier for students or new graduates to get into without any experience. This can be for community service organisations, environmental or ethical campaign groups, research or educational programs, or even political parties. Since volunteering roles are less formal, you may be asked to do tasks that seem menial or less related to your specific degree. For instance, you

Figure 8.7   Benefits of volunteering work.

might spend your time assisting at a local food bank by organising donations, preparing meals, or serving meals to those in need. Alternatively, you could offer your labour by participating in a community clean-up event, where you help clean up a park or beach. You could help with a recycling initiative in the local area or participate in sports volunteering where you coach or carry out organising tasks for your sports club.

Volunteering activities can vary widely for different organisations and may depend on your own time and skills. Here are some common volunteering activities that you may be interested in:

- **Community cleanup:** Participate in local cleanup events to keep public spaces clean and environmentally friendly.
- **Food banks:** Help sort, pack, and distribute food at local food banks or soup kitchens.
- **Homeless shelters:** Volunteer at shelters that provide assistance to homeless individuals and families.
- **Animal shelters:** Assist at animal shelters by walking dogs, cleaning cages, and helping with adoption events.
- **Tutoring/Mentoring:** Provide academic support or mentorship to students, especially those in underprivileged communities.
- **Elderly care:** Spend time with seniors in nursing homes, offering companionship, playing games, or helping with daily activities.
- **Hospital volunteering:** Offer assistance in hospitals by helping with administrative tasks, comforting patients, or supporting medical staff.
- **Environmental conservation:** Participate in tree planting, beach cleanups, or conservation projects to protect the environment.
- **Disaster relief:** Join organisations that respond to natural disasters or crises by providing aid and support to affected communities.
- **Blood drives:** Volunteer at blood donation centres to help with registration, refreshments, or donor care.
- **Community gardens:** Help maintain community gardens or food gardens.
- **Home construction:** Work with organisations like Habitat for Humanity to build or repair homes for those in need.
- **Crisis hotlines:** Volunteer for crisis helplines to provide emotional support and assistance to those in distress.
- **Library literacy programs:** Assist in literacy programs by teaching adults or children how to read or improve their reading skills.
- **Sports coaching:** Volunteer to coach or assist in youth sports programs to promote physical activity and teamwork.
- **Arts and crafts workshops:** Lead or support art and craft activities for children, seniors, or individuals with disabilities.

- **Mentoring entrepreneurs:** Support aspiring entrepreneurs by offering guidance and mentorship in business development.
- **Community events:** Help organise and manage community events, festivals, or fundraisers.
- **Computer literacy programs:** Teach basic computer skills to individuals who may not have access to technology education.
- **Political campaigns:** Volunteer for political campaigns to support candidates or causes you believe in.

These opportunities can provide you with valuable practical lessons and enhance your transferrable skills, especially when you are first starting out in your career. It may be a chance to try new things or use your STEM skills to support groups or causes that share your personal passions. For example, if you have a passion for animals, you may find fulfilment in volunteering at an animal shelter or participating in wildlife conservation efforts. On the other hand, if you enjoy working with children and want to make a positive impact, volunteering as a tutor for underprivileged kids could be a meaningful choice. More general volunteering experiences like preparing food for needy families, helping in a charity's warehouse, helping the elderly make small home repairs, reading to people in hospital, *etc.*, all help develop a sense of responsibility, develop your transferable skills (*e.g.* communication, patience, organisation, *etc.*) and show that you can reliably show up and be a trustworthy team player. These are excellent things to help round out your CV/resume. By pursuing volunteer opportunities that resonate with your passions, you can make a significant difference while also finding personal fulfilment. You will also learn a lot about companies, organisations and how certain industries operate.

Volunteering does not necessarily require a lot of time and can be quite flexible. Sometimes roles are only a morning a week. These roles are another chance to connect and network, as well as learn how organisations operate. Larger groups, like not-for-profit organisations or political parties, often have paid positions and roles for those who demonstrate reliability, teamwork, and relevant key skills. These experiences also stand out on your resume and can lead you to many other new opportunities. You might already be associated with a charity or community group (religious groups count too), so your first step can be to ask where you already have connections. If you are unsure where to start, work through Activity 8.6 to work out how you might start volunteering.

Activity 8.6: Start Volunteering

Activity goal: Find somewhere to volunteer.

Purpose and benefit: Even if you aren't going to volunteer this month, it is useful to work through where you might volunteer and what aspects might help you and build your CV/resume.

Activity steps:

1. List any of your hobbies or extra-curricular activities that you would enjoy spending more time doing, such as sport, reading, gardening, cooking, music, *etc.*

2. Now list any passions or topics you really care about, *e.g.,* the environment, teaching, animals, *etc.* Together the lists from steps 1 and 2 are your starting point for working out good volunteering spaces to look at.

3. Search for your topic + location online or using social media platforms.

4. Contact any group that appeals to you and ask if you can volunteer. If you haven't volunteered before, be assured that these groups are very used to taking on volunteers. It is part of how they function. So, if you are reliable and trustworthy, you should always be welcome.

5. Have your resume ready and be prepared for an interview/chat. They will likely want to talk to you at some point before they work out what task to assign, whether you are a good fit or have the right availability, but this is not like a formal job interview.

6. Be prepared to start with really basic tasks while the staff get to know you and you get to know the needs of the group. This is all part of your contribution, so be patient while you build an understanding of what they do and how they do it and you can move on to more engaging or more tailored work.

7. If your interests change or a plan didn't work out, revisit steps 1–6 to reassess and get further inspiration to try somewhere else.

Takeaway: Volunteering can be very rewarding and for most people it isn't just a resume filler or something you do because your family drags you there. Volunteering can be a feature on your resume/CV if you present it correctly. Remember to highlight what you learnt and how it is relevant to the role you are applying for. Volunteer experience that is just listed with no comment or reference is not that helpful. So also put in the effort to briefly showcase what you learnt that is applicable to the role you are applying for. Remember, this means your volunteering description will also need to be tailored in your CV each time.

Conclusion

This chapter is useful if you are still studying, giving you some time to explore job markets and reflect on industries that could use your skills. However, this is just the beginning. If you have graduated already (and are possibly feeling a bit overwhelmed), these activities could help you better narrow down your search. If you are working and don't feel like you are in the right place, perhaps now is a useful time to recall your priorities and values to better align with your next job search.

Applying for a job takes a lot of time and effort. Most applications take hours or even days to complete, so it's better to be selective and focus on the ones that are the best fit for you. You now know where to find job advertisements and can hopefully recognise the type of job you would like (if you are still struggling to know what you want, re-visit Chapter 2: Transferable Skills and Reflection).

To find the best fit for a job, first think about your qualifications, skills, and interests, and see if they match with the job requirements. Also, consider if the company's mission, values and culture align with your priorities. Remember that your preferences may change with time, and each new job experience will give you further insight into what is truly important to you. Carefully assessing the job description is a useful first step to make sure you spend your time and efforts productively.

In this chapter, we have explored a few ways to find a job, such as online and physical job centres, networking, volunteering, and internships. We also delved into important factors to consider when looking for a job, including salary, working culture, and personal gains.

The job searching process is better un-rushed. Hopefully, you can give yourself time to explore the job market, even if this is during your studies. And there is no right or wrong job. Any path will help you gain valuable experiences and skills to develop your own career. Volunteer where you can, sign up to internships, participate in research projects, engage in networking events, or seek mentorship to further shape your understanding of different jobs within STEM fields.

Extra Considerations When You (Will) Have a Higher Degree by Research

ANGELA ZIEBELL[a] AND RICHARD HUYSMANS[b]

[a]Deakin University, Australia; [b]Health Education Australia Limited, Australia

While this whole book was written to help all STEM students and graduates, we wrote this chapter to address many of the questions that **mainly** impact those that have done (or are still doing) higher research degrees. This chapter will look at how that extra study positions you to move into slightly different roles and we will help you work through what you might want in your career and why. Right at the start of the chapter, we talk about a few issues that might be different for people in your position, why that comes about, and share our thoughts and advice on these points. Then we go through a range of common positions that PhD and research master's graduates frequently head to so that you can understand some of the most common roles. In this chapter, we will call this group HDR (higher degree by research) students/ graduates.

If you have already finished your HDR, or are currently a post-doctoral fellow, this chapter is still for you. Do not worry if this is your first time thinking about your career in a structured way. We've never met someone who regretted starting to think about their career planning and acting early, but we've met plenty of people who started late wish they had sat down and thought about their career earlier. The journey is manageable no matter when you start and no matter what place you are coming from, the journey just can't happen without you.

> "The best time to plant a tree was 20 years ago. The next best time is now."[†]

[†]Chinese proverb.

About the Authors

As we both have non-traditional paths and are HDR graduates in STEM, we thought it might be useful for you to first read about our backgrounds, so you know where we are coming from. This will also give you two examples of non-traditional pathways post-HDR. Throughout the chapter, there are also snippets from other PhD graduates as they share their career journey and insights.

Angela worked at the Australian national laboratory CSIRO as a research assistant before taking on a PhD project at CSIRO. Afterwards, Angela did a post-doc in an area very similar to her PhD at a national lab in the USA, staying to become a staff scientist and project lead, before returning to Australia 5 years later and taking some time at home with her young family. In returning part-time to work, Angela took up work teaching and doing data analysis at the university she had graduated from. Angela then took up a full-time education designer role, merging her lab management and teaching skills with knowledge of how to work with industry, in order to redesign chemistry laboratory classes and link them to chemistry done every day in workplaces all over the world.

After this role, Angela worked as a teaching-focused academic, running the career learning unit/class for a science faculty and the internship/placement units. During this time, Angela continued to build her education research interests as opposed to the chemistry research interests she had previously worked on. Angela's research focus is on the skills STEM students need to prepare for successful employment that they find rewarding, not just to get a job. Angela also has an interest in ensuring that university teaching is supporting all students to reach their goals and that students that need more support (*e.g.*, less family knowledge in this area or needing to work heavy hours while studying) are getting the support they need. While Angela's role is currently as an academic, it was 15 years between PhD and getting back there, and while there was lots of hard work, it was never the end goal. Angela could also see herself working in either industry or government in the future and is not tied to staying in higher education. Angela is more interested in where she can do the best and most interesting work to help with big problems.

Best thing about your unique career path: Being able to do things that no one else seems to have done before because of my unique experiences (*e.g.*, developing an Indigenous Science unit to help STEM students better understand how Indigenous perspectives and science have great value).

Worst thing about your unique career path: It can be a bit lonely if you don't actively go out and find people who look a bit like you and have interesting conversations with them. However, there are always people to find to talk to about your individual interests.

Richard went from bachelor's degree to honours (an intense 9-month program common in some countries, often 50% theory and 50% thesis) to PhD. If there was some kind of study after that – he would have done that too. For Richard, study was the easy path. It required zero decision making. Of course, not making a decision is still a decision. Months before finishing his PhD, he began looking for non-academic jobs. A post-doc (in the traditional sense) was of zero interest. Through that process, he learned (1) a pharmaceutical sales job was not for him (and if you have seen the series called Dopesick you'll know why); and (2) there are many other kinds of ***post-docs***, not just ones involving academic research. Prior to submitting his PhD, he landed at job at Monash University – where he had already spent 8 years – and spent 4 years there as a strategic advisor in areas as diverse as medical training, criminal justice and ageing. This gave Richard a taste of the consulting life, and within 3 months of having the desire to go out on his own – he did. For 14 years, Richard ran his own little consultancy. At first, it was writing tenders and grants, and later it transitioned to career advice, coaching, mentoring and training. But the whole process gave Richard a keen insight into all things business – from the mundane of social media, marketing and sales, to the fundamentals of profit and loss, balance sheets and financial planning.

Like every small business, COVID had a huge impact on Richard. With such a large portion of his business coming from the university sector, it was no surprise that business slowed down. Ultimately, in 2022, Richard made the tough decision to ***join the workforce*** and take on a salaried role with Health Education Australia Limited (HEAL). In this role, Richard uses none of the skills he learnt in his PhD, and not much of what he learnt working for himself. However, this role would not have been an option if not for his PhD and working for himself.

Best thing about your unique career path: All jobs are options. Richard has had many recruiters mention his unique and diverse career path and how they view it as an asset. It also means Richard gets considered for many and varied roles.

Worst things about your unique career path: Being a small business owner is admired outside academia but is poorly regarded in it. And being self-employed means you've got to make the money to continue. In that respect, it's like being an academic. Regularly needing to "apply" for your own job. Demonstrating that I have usable and useful skills has also been hard, as I have now transitioned into the wider workforce. As a consultant, people were happy with a portfolio of projects demonstrating I could do the work. As an employed person, looking for further work, those same projects don't seem to hold the value they did as a consultant. When I was self-employed, I sat on a few recruitment panels. And, at least on one occasion, a self-employed person was interviewed. The panel discounted their self-employment as having any relevance to the role. Despite the requirement for a great deal of autonomy and self-motivation and self-organisation within the role we were recruiting into.

Career Issues That Mostly Impact Higher Degree Research Students/Graduates

Below, we go through a handful of issues that impact many HDR graduates and students. While some undergraduates might have experienced challenges around these points too, most of these challenges are not common enough among undergraduates to include in the general section. You might recognise yourselves or friend's experiences in many of these or you may not. It is very unlikely that you experience all of them, but everyone's experience is different. While reading the following, consider whether you are impacted by these issues or whether there is someone else in your circle that might gain by reading this section. As you go through, there will be some exercises to help you think more deeply about any challenges you have, but also to help you start to plan a way forward as you read through this chapter.

Uneven Sources of Advice

Most academics encourage students that are highly successful in their area to pursue further degrees. To "make the most" of their talent. For undergraduates that do not know if a higher degree is what they want, or those that have just not thought about non-higher degree paths, there can be limited voices that represent all the other career paths. That is little support or encouragement of other pathways from the student's academic educators. This can mean you spend a lot more time thinking about academic careers and little time learning about non-academic careers, of which there is an almost countless array. This is for a number of reasons. When you sit down with those that surround you, like in Figure 9.1, how many different views do you get access too? If the views aren't diverse enough, it's time to access other sources of advice.

It makes sense that academics would encourage and support a movement towards academia – it is what they know. It is what they are expert in. It also makes sense that academics are unlikely to give career advice on how to enter a particular career outside of higher education because

Figure 9.1 It can help to look for advice from lots of different people. Photo by Jason Goodman on Unsplash.

most do not know a lot about jobs outside the university. Giving career advice is also a specific field all by itself, so even those who have experience from outside higher education might not feel they know enough to give advice.

To help you gain more of a balance, you can look for other information and opinions. The rest of the book is full of advice and activities that can help anyone (including you) work through their career thinking. Once you start to access general careers advice at your institution, read about careers and work through some of the activities in the book, you will start to get a more complete picture of where graduates can go after completing a research degree too. You will need to read the same background and then also look for additional information about graduates with additional qualifications. For instance, you will likely have read about graduate programs. Some governments and big companies have graduate programs that have a stream for those with research qualifications.

There are also career pages that focus on helping people with, or completing, research degrees. These can be found by a general search engine search using phrases like career advice for PhD/master's graduates or master's/PhD-career-STEM-graduate. Like any search engine search, you will need to try a combination of words to see what gets you the best result for your specialty and for your geographical region. Also, keep in mind that bigger markets like the USA have a lot more blogs, website, and videos. You can learn a lot from these even if you are not going to work in the USA, but be sure to apply a filter. Understand that they are giving advice for people in the US and your situation could be different. Generally, these videos are great for helping you understand the issues that might be at play. Then local resources and resources in your discipline can help you fine-tune your understanding.

Attachment to Career/Specialty

People who complete a research degree are probably more attached to their discipline than the average graduate. They are spending 3–6 years further pursuing a specialty, after all. The dedication to their discipline is what they need to persist. However, either subconsciously or consciously, some people who go down this path view that, or are taught that, this is a very "pure" path. Careers that follow this path might be suggested to be more ethical or passion-filled than almost any other career choice. In reality, essentially any career can be entered into with great passion and enthusiasm, including careers that others would dislike. Passion for your career is all about finding what you love, not about certain careers being worthy and others not being worthy.

I'm an Expert or I Need to Do It All Syndromes

OK, not real syndromes, but bear with us. Success in higher degrees requires a certain amount of skill and self-motivation, as well as self-education and training. Having achieved your higher degree (or being almost complete) provides graduates with a sense of "I can do anything because I am good at learning". And this is certainly generally the case. As a higher degree graduate, we felt it too, it gives confidence that you can start new things and succeed because you did once before. This can be a wonderful asset. However, this approach can also sometimes be misplaced and has a couple of side effects.

First, the training can lead to a tendency to do everything yourself and rely on yourself for everything. In some environments this is crucial, in others it needs to be toned down. In others still this level of self-reliance might be totally misunderstood. You might be madly working 60+ hours a week, stressed out trying to get everything done and the boss will wonder why on earth you do not talk about what is going on and get help to bring in the other skills they have available. Angela has definitely experienced this. As someone who has worked in teams big and small, with their own team and by themselves, Angela has sometimes realised that she was doing it all just because she had failed to realise that asking for help was an option. There is no magic fix to this other than keeping these points in mind as you reflect (Chapter 2 onwards) on your career and career goals or keeping it in mind as you work. Being aware of a possible problem is the first step in making sure it does not happen or in correctly identifying the problem when it does happen.

People who have pursued very high levels of training often love complicated ideas. This can lead to rejection of advice because it seems too simple, too obvious, or not sophisticated enough. For example, people often ask Richard "what's the evidence for the value of a shorter resume?" expecting his response to be an academic paper. Failing to be able to respond with a reference, the advice (make a shorter resume) is often rejected because the HDR student/graduate wants to believe all of their hard work and effort needs to be documented. They tend to feel a short resume is too simple. It does not tell the whole story. Quick question, do you want to read 25 applications at 30-pages per resume when you help interview? No, neither do we.

> "People who have done PhDs need a certain level of "knowing" they have the answer to back themselves through years of research in a PhD and then potentially years of applying for funding. So being convinced you know everything can have its purpose. There is a tension though between "knowing" you are the right person for the funding and accepting other people also have amazing ideas."
>
> – Michelle, research fellow at a university and director of a research business. Formerly taught writing.

The short resume rocks. No one needs a *Nature* publication to see that. The same protest can be given when people ask about the evidence of resumes with nice aesthetics. Again, there is limited academic research in this area. The one that looks like a research grant. Twelve-point font. Five-millimetre margins all round. The advice can be rejected on the grounds that they know better or there is no evidence they aren't right (ironically). An element of this is it can be difficult to move onto other forms of writing and presentation. So, it is natural for data driven people to ask for data to prove they need to change their approach or learn something new. Thoughts and comments like the "facts should speak for themselves", "My experience should be enough" and "I've completed a PhD, writing a resume is not that hard" are not unusual and have a certain logic. But sometimes others, that do not have more training, have the knowledge that you do not have and that's OK. In fact, it's great. You do not have to have it all under control, you just have to know how to find a few expert opinions and work out what to do. Note: experts aren't always the ones with papers on the topics – that's key to this point.

This leads us to the second point for those whose "go-to" is to do it all themselves. Think about how much you can get done when you do not have to do everything yourself. If you can bring in other ideas, concepts and skills to your work through recruiting others (employment or collaboration) you can do so much more. You can do your work faster or more efficiently; you can apply your work to new areas, you can come up with new specialties by putting two specialties together.

Thus, as you read through this chapter, I ask that you bear all of that in mind. That you realise some of your training has probably conditioned you to reject the simple, or think you are able to do anything on your own or without help. Or you may fail to recognise that seeking help and following the advice of those who do this for a living will get you to your goal faster. To start us off, let's reflect on what we do and don't like about research in Activity 9.1.

Activity 9.1: Reflect on What You Do and Do Not Like About Your Research Work

Activity goal: To stand back and reflect on what you truly like and dislike about your research work and what you are neutral about. This task will take 1–2 hours.

Purpose and benefit: This reflection will allow you a chance to distil down what it is that you might like to keep in future work or avoid in future work. It will also help you build questions you can use at networking, career or industry events to help understand the key features of different roles that are being offered.

Activity steps:

   1. Write a list of at least three things that you like, dislike and are neutral about that are central to your master's/PhD work. These points could be anything. It could relate to your work environment, the equipment, the people, the topic, the narrowness or the size of the project, the uncertainty, *etc.*

   2. Now that you have that list, reflect deeply on at least one of the things you like and one thing that you dislike using what you learnt about reflection in Chapter 2: Transferable Skills and Reflection. What is it exactly that you dislike/like? Why? What does it mean to you that you feel this way? Is there any history with that feeling and what it means to you? Is there a pattern? Is

Activity 9.1: Reflect on What You Do and Do Not Like About Your Research Work – (*Continued*)

it about more than just this one thing you like/dislike? For example, are there confidence issues or fears of failure, a need to appear a particular way, *etc.*

3. What do these insights mean for the career you are looking to move into? Do any of them give you reason to think twice about your plan? If they do, write out 50–100 words articulating why you think you might need to look at a different or certain career path given what you have learnt about what you do, and do not, like about your research work.

4. Picture you are at an information night at a large company that offers multiple roles with your training (or the training you will have at the end of your study). Based on your insights from point 2 above, what questions do you ask the company representatives to work out which type of position you would prefer to apply for? Write five questions.

5. If you have another role (even a casual or non-professional role) repeat 1–4 for that role. Is there much difference? Did you learn anything additional? What was it?

6. Find at least one event to attend to help you explore different roles and see where your skills might be used (this can be online). Go to the event and make sure you ask all your questions you have written at least twice each (not necessarily to the same two people).

Takeaway: What you like or dislike during your research degree can give you a lot of information that will help you plan where you want to go next in your career. Staying on a predictable path is one way forward, but statistically, the chance of this being the best option forward for you is a lottery. Most people will be happier if they think about their career, and branch out and work out where they want to go.

Hyper-focused

Higher research degree study tends to reward students for being hyper-focused. After all, how do you get through a higher research degree without your project being your main focus? This hyper-focus can be a super-power, but if hyper-focus is not something that you are skilled at, the priority put on hyper-focus can lead to thinking you are not right for science or maybe just science research. There are a couple of really important things to note here. Science needs all sorts of brains and all sorts of work approaches. So, however you approach your research or your science, there is a place for you if you want it. Some places are just a little harder to find.

The opposite of hyper-focused, isn't someone who is unorganised or unprofessional, it is someone who wants to look at the big picture and all the interacting issues. This is a very important role and, as science gets more advanced, it gets more complicated, so it is increasingly unlikely that any meaningful project can be done without working across the more and more focused specialties. There are places for people with big picture approaches all over science and the workforce (note not just the scientific workforce). The key to a good team in all workforces is a mix of strengths and skills.

How many people in your lab can look intently at their data like this guy in Figure 9.2 and not realise that you are talking to them? It is true, though, that academia rewards hyper-focus more than potentially any other career avenue. So, if one of the things that you don't like about the academic research pathway is that hyper-focus element, then know that other workplaces do not put such a high priority on focus. In general, workplaces put a lot of focus on you being able to work on completely different projects, even an ability to move teams. This isn't just the case in for-profit businesses either. This can be the case in research teams too, for example, in national laboratories and in companies that refine or design their own product.

If you have been feeling like you don't belong in STEM because you do not want to be hyper-focused, then there are still plenty of avenues for

Figure 9.2 Hyper-focus is a superpower but needs to be understood and used when appropriate. Photo from Nicholas Flor on Unsplash.

you to explore. It is just less likely that a traditional academic role is for you. For those that have been surrounded by academic role models, this just might be hard to see or it might be something that hasn't occurred to you to think about yet.

Not Wanting to Disappoint

A related observation is that research students or new graduates often do not want to "disappoint" their supervisors or others around them that appear to want them to pursue a certain path. This might be parents, grandparents, or peers, but is often a supervisor. Let's take the supervisor first. Family is always trickier.

It can be easy to see your supervisor, whose research is most of their life, as someone who will be disappointed if you do not follow that academic pathway that they followed. There are a few really important points here. First, do you even know that they will be disappointed? Check how they talk about other students that have not followed their path to get a sense of what they really think. Make sure it is not just what you are worried about. Most supervisors are very realistic about the number of academic or research-only roles that exist for graduates.

If you have checked that you have gotten the correct sense of your supervisor's attitude towards pursuing non-academic career paths and it is negative (this won't be a lot of supervisors) then there are only so many conversations you can have with your supervisor about careers and that's a shame. But you need to know that a strong opinion that students should be "good enough to succeed like them" is a very unreasonable stance and very unfair to you and your fellow students. Depending on their age, it also totally ignores changes to the education system that have happened since they have come through, meaning that it is harder to gain a position now.

So, while we understand it can be very unsettling, try not to get too worried about what your supervisor thinks. Maybe a former student from your research group also left to work in a non-academic or non-research focused role and you could contact them and share your frustration with them. This person could be a great mentor of how to manage that situation well. Even if they did not manage the situation that well, they might have tips on what they would do differently next time and why.

If the pressure comes from your family, this is a bit harder. Everyone has their own specific family situation, so it is hard to advise if this is your situation. Finding a mentor who has lived through leaving research post-research degree could be a great help, just as above. The number of people who struggle with balancing what they want to do with their career and family and social expectations is much wider than that, though. You might be able to have great conversations about this issue with people outside of STEM. And this situation also exists for those that chose a degree that family didn't like. This means there are a lot of people in your position if you are struggling with getting family to accept a career or education decision. Maybe you have a high school friend or another family member who has struggled with not wanting to disappoint family. They could help you talk through your priorities and options.

One option should definitely be to consider why your family has certain expectations. Was it their passion, for example? One they did pursue or maybe one they missed out on, and they were hoping to give you a chance they missed out on?

It's important to remember that no one is good at a job that they hate. So, if your problem is how much you are burning out because you do not feel any joy from your work, it is important to think about talking to your family to let them know. Also, academic and research work can be very competitive, and work can be from contract to contract. Most parents are great at working out these sorts of risks and steering their children away from roles that are lower paid or "risky". But if you are thinking about moving away from academia or research due to the unstable nature of the work in your environment, then think whether your family appreciates that point, or whether they need more information to understand the challenges.

Last, is there someone in your family that is a slightly friendly ear? Someone that will keep a confidence. They might have many of the same feelings as the rest of the family, but you can trust them to share what you are worried about. Someone who is worried about you but understands your challenges. If you have such a person, it might be a really good practice conversation where you can work out what you want to do/say next. They might also have specific advice to help you navigate the situation.

There can also be pressure that is either overtly from peers or often from yourself related to what peers are doing and you are not. Everyone has different career goals, so try to keep in mind that someone else's perception of what you will do is nothing more than that. A perception. You don't need to live up to it.

The really important thing to remember is that you are not the odd one out here. A huge number of people leave academia and research each year as they finish their HDR study. And, of course, at all other stages in their careers too. In some countries, this is more or less common, but in most countries there are many more HDR students in STEM than there are new jobs for researchers and academics. AND it is very possible to leave and re-enter through a number of mechanisms (these vary greatly with discipline and country). This chapter looks at both the opportunities if you stay in research and academia and those that are out there if you join government, a business or a not-for-profit.

Wanting to Be Able to Control What You Work on

Some considering whether to leave the university are worried that they will not have freedom to work on what they want outside of academia. However, it is important to think about the reality of the situation. Most people don't choose their PhD or master's project, although they might strongly influence it. This is akin to the sort of roles that you might have outside of the university. You obviously pick which role you apply for and then once you have a certain level of experience you will likely decide things like what techniques you use on a given project, who you reach out to for consultation or to work with, who you hire, *etc*. Likewise, those of us who continue to post-doctoral fellow only have so much control over their research. You will only be able to apply for positions that exist, competition is usually fierce and once in the role the senior academic will still make major decisions and may also dictate how many students you need to help with, or if you get assigned teaching, which may not be clear in your contract, *etc.*

In all workplaces, a great boss, a great project and an ability to work at a level of independence you thrive on is key to enjoying your career and to progressing. But none of these things depend on you being either inside or outside of a university.

"As someone with a lot of years of experience (not in STEM) and multiple qualifications who has worked with numerous newer PhD graduates, I'd suggest PhD graduates ensure they recognise other people's life experience and the skills that they have developed in non-STEM sectors. It's incredibly galling when a junior PhD graduate won't recognise the value of other skill-sets and experiences and it can be extremely detrimental to the workplace. Making yourself easy to work with in a team is an incredibly important skill at any level."

– Alix, political science graduate with a law master's and two decades of experience in various areas of government and not-for-profit organisations, specialising in policy.

But if I'm a fully-fledged academic it will be different, right? Yes, and no. First, you have to land a role. In many countries, this will be a temporary role initially and could be after 3–8 years of post-doctoral fellowships. An ongoing position will often take 3–5 years if hired on a non-permanent basis to start with, although this time might be might much higher or not needed, so check what is normal wherever you are interested in working. Now your research needs to be funded. So you need to propose projects that you think are great, that you can do, that have the highest chance of getting funded. Yes, you decide what goes forward in your application, but not everything attracts equal interest, so it is normal for people to have to follow one particular path of their interest because that coincides with success in funding.

However, competition is fierce no matter which environment you are applying in. For those who are unsuccessful achieving a minimum amount of funding (different everywhere), roles will either end or you will be moved to focus more on tasks that help run the university, like more teaching and administration (*e.g.*, coordinating a large degree program, supervising the research student administration for the school, being the academic safety lead, being part of an ethics committee, *etc.*). These roles can be extremely satisfying roles and you will use a great array of your transferable skills while learning new technical skills and understanding the university as a system better. But the truth is, an academic career is far more complicated than achieving that first academic position. There are a large number of things that decide what you work on, just like other jobs.

Now you have a few new insights into what you might want out of a career post-research degree, and you hopefully better understand some of your motivations. Let's next build a career wish list in Activity 9.2 and then look at some example positions that PhD and master's graduates go into from a big range of STEM disciplines, maybe all STEM disciplines.

Activity 9.2: What Is Your Career Wish List?

Activity goal: Work out what you want on your career wish list.

Purpose and benefit: Get a clearer picture of what are your hard limits on what you want to have, and avoid, in your career. Having a list also means that you have a clear picture of what you want and allows you to check if your expectations and desires are realistic or not. Or how you make them realistic. For example, now that you have written them out and reflected, is one more a "good to have" rather than a "necessary", leaving you more room to concentrate on the other points?

**Activity 9.2: What Is Your Career Wish List?
– (Continued)**

Activity steps:

1. Write down three things that are "must haves" for your career. Maybe these aren't achievable at your first workplace after graduation, but there is a clear line to them in a few years. There is no right or wrong answer here and it is important to note that your "must haves" will change over time. This could be because you are in a different life stage, the industry has changed, you have grown out of a particular goal through experience, the "must have" wasn't what you thought it was, *etc.*

2. Now do the same thing with three "must minimise" points. Note that a point can be on a person's "must haves" at one time and on their avoid list at another stage in their career. This is very normal, so it is useful to revisit this activity each year or more often if your life is changing quickly. For example, travel is a popular goal for young people but becomes very difficult once you have family responsibilities. Employer supported study is great but needs to be timed around years when you have free time in your life.

3. Reflect on whether your career thoughts or plans take these "must haves" and "must minimise" points into account. If they are not accounted for in your plans, how might your career plans need to change as they develop to take account of these "must haves"? Once you get to this point, you might start to realign or rethink what you really want in your career and what you want to avoid. If that is the case, try this activity from the start again now that you have given all parts some thought.

4. Talk to (or email) two experienced professionals, using your best networking skills, and ask them what their must haves were when they started their careers and why. They may not have formally had any at the time, so you need to ask what they would have been if they were asked this question at your point in their career. How did their must haves change so far in their career and why?

5. How has the discussion in point 4 influenced your job wish list since you first started at point 1?

Takeaway: Often when we have ideas about our future career, even really clear visions may not have been put together in a structured manner where the person considers all angles and puts in place a plan. Careers (without career learning like this book or great career education classes) are often planned around roles we connect with, a personal interest, and roles that we have seen in the community. By occasionally stepping back and analysing your thoughts like in this activity, you can bring in new information to the process of career planning, making it more likely that you will get to a highly satisfying career *via* a more direct path.

Some Example Positions for PhD/Master's Graduates

As I hope we have made clear, STEM research degree graduates can go into roles essentially anywhere. But to help you out with a few basic descriptions that are common across disciplines, here are a handful of role descriptions. Please do not let the presence of these common descriptions stop you from exploring other titles and roles. There are literally hundreds. Which is exactly why we did not try to put them all here. Note that the need to have a PhD or master's qualification for different positions varies hugely around the world and even across disciplines. For example, there is a lower percentage of engineering students who go on to PhDs, so engineering positions are less likely to be filled by someone with a PhD in most countries as there are a much lower number of these graduates. Countries that have a lower rate of graduates with higher degrees are also less likely to have PhD graduates in any given position.

Some of the following roles have research as central to the role, others use research skills in a repurposed manner. Data from Australia suggest the majority (80%) of research roles are actually outside universities, which makes sense. Research is needed everywhere, not just in universities. Most departments in medium to large companies have some kind of analyst, strategist, investigator, or developer whose role is to look at data, develop options, and suggest ways forward. Beating a path where there previously was none. These might not loudly shout research, but they use many, if not all, of the skills that a more stereotypical research role uses. They are roles that many PhD or master's graduates can thrive in. For example, we know of researchers who completed their degrees, master's/PhDs, and postdocs in science, and then moved into analysis roles in energy companies. Not to analyse energy use, production, or distribution, but to analyse customer payment data.

We know of people who completed physics degrees, master's/PhDs and post-doctoral positions and then go into roles in banking and finance. These people research and develop new products and services across the spectrum of personal and business banking within the bank's research and development team. Opportunities to use skills from your STEM degree in another context could be in IT (the T in STEM), government, insurance, consulting, pharma, biotech, utilities, manufacturing, farming, trade, agriculture – the list is endless.

For those who don't have a central research focus, we still see research skills used regularly. The HDR student manager who uses their research skills to keep up with best practice internationally for their field, the government employee who works on big data projects and sources software and techniques for complicated analysis, the research officer who has to stay across all the engineering fields in order to help write grants, *etc.* So, you will (hopefully) see that research can be present in a huge range of rewarding work. Just in different forms. We hope these explanations of roles help you when you are trying to navigate the job advertisements, company websites and networking sessions.

Government Employee (Non-research)

It is often assumed by HDR graduates that there are few routes into government roles post research. However, this often is not true. Governments hire people from all backgrounds for both their specialty knowledge and for their transferable skills. If you are still a bit unclear on how your transferable skills might work in a government context, we suggest you revisit the transferable skills section of Chapter 2 and review with a government role in mind. In addition, although they can be a little hard to find, always look for recruiting programs that either take applicants with a higher degree by research or that are specially tailored to such candidates. In my experience, this may take some phones calls and extended internet searches. If this is a possible interest area for you, it is well worth your effort. Programs like this usually take people as they are finishing up (*i.e.*, you can apply before you hand in your thesis) or in the following two years after you pass your research qualification (however, always check because these times can vary). The institution you graduated from can usually help with finding out any of the above information. Having said that, what might you do in government?

A government employee who has higher degree training in STEM can work inside the government in a huge number of ways. Exactly how common STEM professionals are in different parts of the government will depend on the country, and it is highly variable, so it will pay to look into the country, or countries, you want to work in. However, do not assume that a lack of STEM professionals in government means you should not try to work in government. This might just be a sign that your skills are desperately needed. Just always keep in mind your local work environment.

Scientists, engineers, and technical specialists consult on any sort of science and technology throughout the layers of government, from the impact of taxes changes to improvements in environmental requirements, management of land, use of satellites, *etc.* Some examples of titles are project officer, manager, advisor, clinician, assistant, clerk, director and administrator, which come after technical descriptions that indicate the field of work or the discipline. Many have coded acronyms like APS for the Australian public service and grades or numbers which denote seniority. This can all be a bit overwhelming to start with. There are usually charts for pay grades and sometimes experience levels that can help you understand what you are reading better. Someone with a PhD/master's might be unlikely to apply for an assistant role but job description terms also combine so there are assistant managers, project administrators, clinical assistants, *etc.*

While some of these specialists might also lie outside of government and consult to government, some are permanently employed by the government and work more broadly and on longer projects. Picture the transition to a new UN-backed globally harmonised chemistry safety labelling system. This needs a team in place for a decade, helping to build policy and legislation. Or a large freeway development. Most likely this will be put out to tender and run by a private company. But there needs to be someone in government that can help manage the start up and running of that process. These STEM professionals will work on a large range of projects and are using their transferable skills alongside their technical skills.

Other STEM professionals go into government and specialise by learning how to write policy or legislation. Without these people in government, policy on important science and technology issues (*e.g.*, climate change, fuel efficiency standards, data security, border control, *etc.*) gets written without any scientific insight. This is a very significant problem and one that is common in some governments where STEM graduates are either not recruited or not interested in being recruited into government.

Finally, a skilled researcher has so many transferable skills that, if the type of work suits you, you could work in any office that needed someone with advanced information finding and collation work (aka research), an ability to work relatively independently, good presentation and communication skills and an ability to work well across specialties, sharing data and resources for the benefit of the project. Most HDR graduates have the majority of these skills and would be able to further develop them in the first couple of years of employment.

Teaching

If you really enjoyed your teaching experience while being a HDR student, there are a large number of teaching roles outside of academia. Teaching in a school usually (but not always) requires additional qualifications. Teaching in vocational training might require a qualification, but these are often quite small qualifications aimed at ensuring you have a basic understanding of teaching theory and practice. Whereas it is common for a schoolteacher to need a Master of Teaching if they haven't done a teaching specific degree. In addition, tutoring can be a flexible way to gain experience teaching or to add income to another role.

Teaching can include the production of training resources, building of micro-credentials, in-house training or the building of such in-house training (online and in person), which are vital to professions like engineering and medicine where keeping professional development hours is crucial to staying employed. So, teaching could be well outside schools, universities, and colleges. You can also work for a training company or as a trainer and educational designer within a bigger company that does many things.

As you can probably tell from this description, some of these "teaching" opportunities are informal, like recording video lessons hosted online somewhere (*e.g.*, YouTube, Udemy, Thinkific, there is a long list of options). If you are going to go in this direction, you will want to be clear on what your budget, incomes goals, *etc.* are. This might be a good time to jump to the end of this section and read about mixed roles to help you balance your budget and career. Networking and communication are key when it comes to getting the word out that you are starting work like this.

Other more formal opportunities are available where many markets have an increasing need for professionals that understand both the learning and teaching side of a particular topic and the technical side. This includes a need for people who are also capable of putting together learning and teaching material online for either synchronous, live, or mixed delivery of online teaching resources.

If you have even better digital skills or want to learn, there is an increasing need for people to build standalone education applications and platforms. These might be schools or for educational play, *i.e.*, gamification of learning. These roles are the type that you can often take from anywhere in the world, increasing their attraction to those that want to stay put or live somewhere where their income goes further than in the parent company's home country.

Teaching can take any number of other forms and is also adjacent to mentoring, an important skill that you can hope to build over the first decade of your career. Teaching usually leads to significant improvement in oral and written communication, not to mention patience. So, remember that whatever role you go into in the future, the skills you may have learnt teaching while you did your HDR will earn you priceless skills that you carry with you throughout your career.

Academic

In most places in the world, being an academic involves both teaching and research. Teaching will include lectures and/or tutorials, supervising undergraduate projects and your research degree students. Research will be led by you where you can find funding – after all your research degree proved you're an expert in your thing.

With leading your own research program comes additional responsibilities. At the very least, there will be administration to add to the list of teaching and research. And that administration can be a varied list depending on the kinds of support you are offered. For example, having an administrative assistant, research assistant or post-doc can reduce or even remove certain parts of your administration. See below for descriptions of these roles and what they might help with. However, you have to be quite senior to get significant amounts of help. Most help will need to come from grants that you successfully apply for or consultancies. With a team comes people and budget management. So, if you plan to have a research group (large or small) prepare yourself to be a good reader and manager of budgets and people.

Post-doctoral Researcher

Post-doctoral positions – aka post-docs – are short term, usually in line with a particular grant. In this sort of position, you can learn new technical skills by working under someone while increasing your leadership skills and getting better at finding funding and publishing. Historically, individuals looking to become academics might have completed one or two post-docs before becoming a tenured academic. More recently, because there are so many PhD graduates in most locations and fields, it is more common for PhD graduates to complete multiple post-doctoral positions and to be faced with the need to consider work outside of that line of research, or out of research, in order to gain more secure work.

A post-doc can help with more of the dedicated research work. Like the research assistant, they can do the experiments, as well as collate, analyse and report on data. They'll also help with writing peer reviewed articles and grants. They are a lot like research scientist positions and often people will progress from a post-doc to a research scientist position where they are available (not necessarily at the same organisation, though). This is what happened to Angela. Post-docs also often help supervise students at different levels and might fill in for their supervisor on certain duties like giving a talk if the supervisor has a clash in schedule or teaching (either on a certain schedule or as a one-off guest).

Research Scientist

A research scientist works in any number of places and runs research in a particular area, usually within a team that might be big or small. A bigger team is usually led by a senior research scientist, someone with a PhD usually, plus more than a decade of research experience. Some of the experience is likely to have been as a post-doctoral fellow, but it does not have to have been. Research scientists work for companies and national laboratories (largely government funded), sometimes big public serving institutions like hospitals, and for governments in their respective departments and institutes. With the right experience and opportunity, you can become a research scientist after being a research assistant in most environments. Further study helps with this but does not guarantee it and is not always necessary.

Research scientists are more likely to work unpaid hours, but many who choose these roles would argue that generally it is worth it, even if they are standing out in the sub-freezing weather for hours like these weather researchers in Figure 9.3. However, others do not agree. The hours worked also depend enormously on where you are working, as what a culture or industry sees as "long" hours varies hugely.

Research Assistant

A research assistant can work at a large range of levels. At entry level, you will do relatively repetitive tasks and keep a comprehensive logbook so that all your work is recorded. You are likely to be responsible for maintaining equipment too. As you get better at this, your seniority can go up.

Figure 9.3 Research science can be a hugely rewarding career, but it is usually very demanding. Photo from NOAA on Unsplash.

You do not run your own project as a research assistant, but an experienced research assistant certainly has the chance to place and organise their own work, put their observations and insights to their supervisor. This includes helping manage group-wide or laboratory-wide issues like ordering, training, equipment maintenance scheduling or safety.

A lot of people (Richard included when he was making these decisions) feel doing a PhD **only to become a research assistant** is a pointless exercise. And, if we were to put the minimum entry requirements for being a post-doc and a research assistant side-by-side, we might find a PhD is beyond what is required as a research assistant. In that sense, having a PhD makes you "over trained" for a lot of roles. They can still be great roles that suit you.

Some research assistants can have specialties that (if you had done a research degree) would likely align. So, you might still be using many of your PhD/master's training-related skills and be in demand for a particular type of analysis or maintenance of systems.

Research assistants tend to have better work/life balance. Most roles are protected from the expectation to do extra work to put out publications, build patents, or put in long hours to meet specific deadlines.

University Administration – Overview

I acknowledge that when most people indicate they would like a job at a university, they mean as a researcher. However, there are also many other very important roles at universities that are not as obvious to students. There are a lot of these positions as it takes a lot of administrative staff to support the students and academics. There are also some advantages. For example, as an administrative staff member, you are more likely to go straight onto a permanent contract or at least get there more rapidly. If it interests you, you still get to be *academic-adjacent*, one of the key things that Richard wanted when he first graduated from his PhD.

The administrative work could be business (as in the business of university), teaching or research related, whichever takes your interest and suits your skill set or a skill set you want to develop. In some places, a research degree is considered useful and necessary to occupy many of these roles. So, the pool of competitors might be limited to other research degree graduates, not the entire pool of degree graduates. Added to this, the fact you have a master's or a PhD is more important than the area it is in. This is because you have to understand the processes and approaches to the research, not the theories or results themselves. So, places like the graduate research office, careers office, chancellery and research office seek master's or PhD graduates from a range of disciplines. This means there is a bigger range of people potentially hiring, not just the type of school/faculty you graduated from.

All the following non-academic university roles can exist at the Department, School, Faculty and University level. The number of people required will vary depending on the number of grants or HDR students the unit (University, Faculty, School, Department) has or wants to have.

Research Office

Some of the most sought-after non-academic university roles are in the research office. The administrative division is responsible for supporting the conduct of research. Generally, the activities are divided into two – pre-award and post-award. This separates the work and roles into people who help researchers before they win a grant – with things like grant identification, grant writing, and finding collaborators. In many cases, people develop specialisations in the pre-award team, becoming responsible for a specific grant or sub-set of grants. Often, these grants are larger value and more prestigious. Thus, success means a lot to both the university and the applicants. The post-award team takes care of things once a grant is won. They help researchers write reports, remind them of due dates, set up collaboration agreements, help account for funds spent and communicate with the awarding agency on your behalf where required. Similar to the pre-award team, members of the post-award team often specialise in large grants from specific agencies that are prestigious. For a university, there is nothing worse than winning a grant only to have it stripped away because you could not administrate it adequately.

The research office might also be responsible for collating all the publications in a year and reporting those figures to the relevant national research agency. In Europe and Australia, there are also national benchmarking activities – people in the research office are responsible for finding the relevant data and metrics that feed into the reports that form a large part of those benchmarking activities.

Student Support

Outside the research office, there are also roles working more closely with higher-degree research students. These could be Graduate School Coordinators, Careers Advisors, or Research Librarians. Graduate School Coordinators are increasingly in high demand. As the professionalisation of research degrees has taken place, there has been a push to provide improved support for HDR students. Graduate School Coordinators might do things like engage speakers or advisors to present to the cohort of honours, master's and PhD students, deliver curriculum themselves, and provide pastoral care for students.

Recently, we have seen these kinds of roles spring up in departments and schools where their leaders see the need to provide more dedicated and specific career support for their HDR and post-doc cohorts. In these cases, the roles could be for people who have a discipline match or similarity, or they can also be for people with HR qualifications.

Business Development

Business Development roles within a university are responsible for identifying, fostering, and managing partnerships with industrial and investor organisations. They provide guidance in structuring collaborations outside the university. This could involve connecting companies with experts within the university to provide consulting or advice, or even developing research projects that help solve business problems. For example, a local wastewater treatment plant may want to monitor their water quality and request the university to use their equipment and laboratories for more detailed analysis. Or a company might want to see if there is a way to turn their waste biomass into packaging.

Collaborations could also lead to joint research projects or even dedicated research labs, which are often co-funded by industry partners and/ or government support. Examples of long-term partnerships between universities and industrial partners include the Caltech Boeing Strategic Agreement, the University of Cincinnati joint venture with Procter & Gamble, and the Dyson Centre for Engineering Design located at the University of Cambridge. These partnerships sometimes lead to the creation of new research centres, such as the Hyundai Motor Group Innovation Centre located at Singapore's Nanyang Technological University.

> *"Remember, few people understand what goes into a PhD. Break it down for them when you are applying for work by talking about what your transferrable skills are that you gained from the time spent doing that research. It can be very helpful to be able to articulate why you did a PhD and what you have learnt (big picture – not your research project) and how that will be useful in the sector you want to go into. Try to make sure that your possible future employer understands that you do know how to function in their workplace. What are your workplace skills? How will you function in a team, etc.? This can be work at McDonalds, put it on your resume. If I see an applicant with just their research and education on their resume, I get very worried as to whether they will cope in a non-research workplace."*
>
> – Alix, political science graduate with a law master's and two decades of experience in various areas of government and not-for-profit organisations, specialising in policy.

Business developers also help find and apply for project funding, as well as supporting grant applications and proposals. They can also guide entrepreneurs from the university through the commercialisation process. This could include legal support, such as negotiating licensing, structuring spin-off companies, and guiding researchers through intellectual property and ownership issues. Business Development roles within a university are critical to building successful partnerships throughout the community, commercialising research, and creating opportunities for innovation and collaboration across industries. Such business development skills are in high demand as these build on communication, maths, critical thinking, and creative abilities. Experience in these roles can lead to many career opportunities in diverse fields.

Mixing Roles

There are people within universities who regularly and frequently consult to industry, providing their expertise to one or more non-academic partners. This can be a useful way of trialling the non-academic environment without leaving the comfort of your current role. It can also be a great step forward in expanding your network.

Further still, there are plenty of people who have more than one part-time job. While this sometimes comes about through insecurity of one or more part-time roles, there can be some definite benefits from having two different roles. You build skills in two different areas. Busy times in one role likely do not coincide with busy times at the second role, and you are more likely to be in a position to try something out. You do not have to leave your current role to try a new one. You can have the security of knowing that one role is ongoing and you are comfortable with it, dropping the other role to try something new.

Some find that two part-time roles, or one ongoing role and one casual role, let them balance their life better (although issues of benefits coverage like insurance and annual leave can be harder to navigate in some countries while working part time, *e.g.*, the USA). Hours can be adjusted dependent on what is going on in your life that year. Of course, there is the risk that two unrelated roles will eventually clash. It can be harder to book leave and you might be worried that working part time means that you are not viewed as taking your role seriously. A possible but wholly unfair opinion. Complete Activity 9.3 to help you find some non-academic or non-research roles for you to apply for.

Activity 9.3: Find Non-academic/Research Focused Roles to Apply for

Activity goal: For those of you who are interested in exploring a non-academic or research focused role, you will explore the different aspects of two roles to familiarise yourself with other position types.

Purpose and benefit: Familiarising yourself with the requirements and features of a range of positions can help you work through what roles you might be interested in and how to best prepare yourself for applying to those roles.

Activity steps:

1. Find a position from a jobs search or website that is not a research-focused role that you want to learn more about, or you think you might want to apply for. This is largely a learning exercise, so if it isn't a very good fit (e.g., wrong location or out of date) that doesn't matter.

2. Make a list of the key selection criteria as covered in Chapter 7: Key Selection Criteria and Interviews. Assess how competent and how confident you feel in each of these using a simple 1–5 scale.

3. Look at the role and work out where you could upskill in a relatively quick and affordable manner. For example, you might brush up on a particular software, refresh a spoken language, reread notes from an area you haven't worked in for a few years, take a free online course about a topic like policy, IP or commercial awareness. If you happen to fit all the selection criteria well, then great. But what does that tell you? It means that you might be the perfect candidate, but it also might mean that you are aiming a bit low. You might be better off applying for something that asks more of you, as it is very rare to find someone hired to a role where they clearly do everything already. So, does that mean you want to pick another position description that's a bit more senior and analyse that?

4. What have you found out about how you suit the role you chose? How did you suit the key selection criteria? Was there any easy upskilling you could do to make yourself more competitive? What does this tell you about this role and yourself? Use reflection to think about these points and then write a reflection (use instructions in Chapter 2 if needed) about your insights.

5. Using your insights, pick another role (it can be completely different or similar to the other one but fine-tuned from your reflection). Repeat steps 1–4 on this new role.

6. Set a reminder to come back in a few weeks (less if you are actively applying for roles) and see if anything in your thinking has changed and reflect on why that is. Why has it changed? What does it mean? Where to from here?

Activity 9.3: Find Non-academic/Research Focused Roles to Apply for – (*Continued*)

7. Repeat as many times as you want while you explore different roles. If possible, try to organise to talk to people in similar roles. Chapter 4: Networking can help with that.

Takeaway: You can learn a lot about roles by thinking about what is required, why, and how that would fit in with your experiences and desires for your career. Running through this process a few times will help you familiarise yourself with some new types of roles, which you can explore using self-directed learning, networking, and reading.

Being a Successful Researcher

Funding

From a career perspective, early in your academic life (master's or PhD student), the number of grants is likely to be more important than the total value. At this stage, you should be trying to get small grants (*e.g.*, conference attendance, travel and small equipment). These grants will be equivalent to something small, like income from a few days' work up to a couple of weeks' worth of income, and are often offered through the department, school, faculty, university or professional society you are part of. If you are unsure about what is required of you or where you can find these smaller grants, ask around. Ask your peers. Ask your supervisor. Ask your collaborators. Don't forget to seek help from the grants office as well as the graduate research school and whoever is the higher-degree research coordinator at the department, school, faculty or university level. It is the job of the latter groups of people to help you in your HDR journey. If the thought of looking at your finance systems output (*e.g.*, Figure 9.4) worries or annoys you, it's time to think about the importance of getting some finance management skills. Most systems are clunky and can be difficult when not used frequently, but they are not designed for finance majors. You can do this.

Then, as you near submission and graduation, start looking for larger grants. Grants that bring you closer to the amount deemed useful or successful in your area (highly variable and dependent on the expenses of running the work). Once again, to find out what is valuable or useful, talk to peers, collaborators, mentors, HDR coordinators, the graduate school and the research office. At this stage in your career (the weeks

Figure 9.4   Managing to bring money in is hard, managing it while it's in is also tricky. Photo from Carlos Muza on Unsplash.

and months before and after thesis submission), all of those people have a responsibility to help you succeed.

There is an increasing focus on self-support as you progress in your academic life. Thus, grant value needs to cover the salary of anyone you wish to hire, consumables, equipment and specialty services like external testing or custom manufacturing. The sum needed for funding varies from country to country due to costs. But it also varies from system to system based on different norms and traditions (*e.g.,* you might be expected to bring in money to cover your own wage, part of your wage or none of your wage). Most systems have a structure whereby people have time to progress before expectations of income go up. The first step can be the post-doctoral fellow. Be sure to talk to those at different levels of academia in any country you want to work in to get a detailed knowledge of the system over time. These systems are complex and highly variable. Finding a friendly post-doctoral fellow to talk to if you are a student can be a great help. Even better, talk to a few and you will likely gain different insights from each.

Because PhD students, post-doctoral fellows and academics often travel a lot for their positions, if you want to learn about another country it is likely that you can find someone who has either come through that system or worked there before. Asking your supervisor for a good person to talk to about a particular country can help if you are stuck. Your supervisor will have worked with most of the staff for a long time and is likely to know where they have worked if they do not have knowledge of the country themselves. This is a good chance to set up a formal or informal mentoring relationship with someone else in your school or that might be connected to your supervisor but at another institution. Not sure you have the confidence to ask for help? It is very common for academics to have formal and informal mentoring relationships with a large number of students and graduates, as well as colleagues who are junior to them. So, mentoring isn't unusual. Also, if you are at a university, think about the fact that most academics are there because they also like teaching (not always the case) so understand that a chance to help by sharing their knowledge with you is a natural thing for most academics to want to do. So, ask someone if you can have 5–10 minutes of their time, offer to buy them or bring them their favourite beverage, find a chance to get a few minutes with them at a function, *etc.* Be sure to look at the networking chapter (Chapter 4) to help you with ideas to build confidence (if needed) and prepare so you look like the amazing budding professional you are when you talk to them.

Ideally, you can find a position where you can get up to speed with securing funding across the first couple of years. But the more you show you can bring in some small income and develop partnerships that are likely to bring in income in the future, the more you are likely to secure a great position. Each little bit adds up. Going for a $100 000 grant is not ten times the amount of work as the $10 000 grant. So, consider this factor when targeting grants or other sources of income.

In some settings, the type of funding is also important. Historically, the most prestigious type of grant to apply for (and win) is the competitive, peer reviewed grant. And, for some academics, obtaining a certain number of these grants is part of their performance metrics – so, even if you are fully funded, if none are competitively sourced you might not have a contract. This applies for more prestigious institutes but is another dimension to keep in mind. Consultancy *vs.* grant income can also be viewed differently. While a consultancy project might mean you were so well-known in your area someone came in search of you, because it is non-competitive, non-peer reviewed, this income can be viewed differently. For example, it is likely to not be viewed as research income, which is often a fair assessment. But at some institutions, only research income is seen as "real" income. Some funding can also be open, but reviewed by a selection panel – therefore competitive but not peer reviewed.

From government tenders, to industry consultancy, to fee-for-service jobs, to licensing fees for intellectual property or perhaps even philanthropy or bequest from an individual, there is a huge range of non-peer reviewed funding avenues. As a recent or soon to be research degree graduate looking to establish an academic career, I strongly encourage you to be open to all sources of funding, but keep in mind the context of both your current and likely future institutions. As long as you can raise the capital you need, spend it appropriately and spend it accurately, where it comes from is only one consideration. It will all help to move your research forward. Different funding sources – particularly tenders or collaborations with industry – will be opportunities to trial those kinds of roles. You will get a feel for how the tendering or industry collaborators operate, what you like about them, and what you dislike.

Publishing

Publish or perish is not a new term. Its origins date back to the 1970s or earlier. But, when the phrase was first coined, the volume of publications and the impact of perish were different. Across the world, across all researchers, the average number of publications per researcher is less than one per year. So, if you're doing more than one per year, you're doing well. But of course, you do want to do well; you want to succeed. You can use the field weighted citation index to help work out where you stand compared to other researchers. A process called benchmarking that can really help centre yourself in your field rather than worry about unrealistic goals. This is important as different fields cite to very different extents. For those of us who want to fill the shelves with our publications like Figure 9.5, keep in mind that quality is better than volume.

Figure 9.5   Publishing is central for succeeding in an academic career. Photo from Chuttersnap on Unsplash.

However, the expectation of performance is much greater than one per year, varying extensively by institute, discipline and the phase of your career. A useful measure is cumulative impact factor and a useful way to work out how you compare to others is to benchmark yourself compared to your peers (for now) or those at the level above you (for planning your future path).

On top of the number of publications, there are key performance indicators (KPIs) that relate to the type and location/publication house/publisher. Let me explain.

The type of publication is relatively straightforward – a book, a book chapter, a journal article, a poster presentation, an oral presentation, a blog, a review, a report, a news article, an opinion piece, *etc*. Then, there is the added factor of peer reviewed, invited, or self-published. Peer reviewed is the gold standard. It involves a review of the publication by two independent reviewers, maybe even three or more. There are a mix of approaches, including anything from double-blind (neither author knows reviewer and *vice versa*) to fully transparent (all authors, reviews and comments known to all parties). If you were invited to submit your article, it demonstrates some amount of kudos, track record or assigned expertise. Self-published shows you took initiative. There are an increasing number of pre-print servers that fall into this category. As do blogs and social media, both of which are often derided by the **academy** as being of little academic value. However, self-publishing, and specifically blogs and social media, have a lot of career and scientific benefits. For starters, blogs and social media can help build a following. And remember that funding can come from anywhere, so having a public following is a useful asset for an academic.

Authorship is not a weighted amount, it is purely "was my name on a paper?". So, for example, if someone and three friends got together and said we'll all focus on a paper each and then share the authorship, you could reach the goal. Just an example, by no means a recommendation.

Within peer review, there are a number of calculated metrics, which are useful to understand, *e.g.*, impact factors, field-weighted averages and *H*-indexes. Usually, it is a combination of these measures that you can use to communicate your accomplishments when you need to (applying for new roles, annual reviews or at promotion time).

Impact factors were popular in the 1990s and early 2000s. The impact factor is the number of citations per article per year for a given journal for the last two full years. Impact factors vary incredibly as they depend on how many people are working in a given area, citing the research. This includes all researchers in an area, so areas with more research spending (*e.g.*, medical research or hot topics) tend to have very high impact factors. In other fields (*e.g.*, STEM education research) most researchers are teaching a lot and have less time on their research and less funding. Research is also almost never done outside an institution, so there are no national laboratories or hospitals doing this research, so there is a much smaller audience to read and then cite your amazing work. The language of publication also makes a huge difference as it controls your readership. Impact factors tend to favour the status quo too – highly read journals are more like to have their articles cited so they continue to have high impact factors and thus get more submissions and can publish more articles.

The citation rate of your work is also important to consider. If a journal positions your work in front of just the right people, you are more likely to get a lot of citations, which is how the H-index is calculated. I will leave you to search the actual algorithms used for the different H-indexes, but more articles with high citations will keep your H-index improving. Review articles tend to get more citations than those presenting new data, thus raising the impact factor. However, reviews can take a long time, so doing a review just for the citations is not a great way forward. Making sure you publish your carefully curated review of the existing data, however, is usually well worth the effort.

More recently, individuals and universities have started using other metrics to try to measure the relative importance of a journal and a specific article. There are even entire fields of research dedicated to better understanding and making these metrics. Field-weighted-average is one such metric. It aims to take into account the citation average of the field or journal and compares it to the average of the article in question. Therefore, a value higher than 1 means your article is cited more frequently than the average in the field or journal. A value less than one and your article is cited less than the average.

In summary, there are no blanket rules, but each area will have its good, excellent and not so good journals. Often, getting a well-matched publication into a good journal is a better option than submitting a paper to an excellent journal where it does not quite fit and getting it knocked back.

You should also be aware of predatory journals. While a specific list would be different for each discipline, it is good to know that there are some journals that are considered junk that are trying to find authors and reviewers that they essentially just make money off (usually through charging for publication of the articles). Referred to as predatory journals, it isn't usually black and white exactly which journals fit in this category, and you can find people with a range of opinions on which journals are problematic. Some are just not fabulous journals, but maybe they serve a purpose for students publishing. Others are clearly terrible and no reputable academic supports these. As always, the best path forward is to ask what is what in your discipline, and talk to those around you to develop your own knowledge. You can also find resources through internet searchers on predatory journals, including lists that might mark things as safe, not safe and worth significant consideration before use. If you are looking for ideas of where to publish complete Activity 9.4 for some fresh ideas.

Activity 9.4: Find Journals You Might Use

Activity goal: Identify two journals to publish your work in. This task will take 2–3 hours.

Purpose and benefit: Through this activity, you will discover some candidate journals that suit your area of research for your specific environment (country, discipline and research focus). This will better place you to make decisions about where you submit your research to, and help you have a better understanding of what different journals want when you start your next piece of research. This can help you include within your research aspects that will help set your research up better for acceptance into a given journal. For instance, a journal might only take a study that has worked across different sites (environmental), cohorts (medicine and education) or classes of compounds (chemistry). You might need a certain sample size or length of data. Or what you read in the journal descriptions might prompt you to go back to your work and determine a few more things about your research before you move forward (*e.g.*, is the theoretical basis clear, is the research gap you are filling clear, how does my scale of work fit with the different journal research scopes, *etc.*).

Activity steps:

1. Using your preferred search engine or database, search publications for authors with an affiliation with your organisation. Get specific in terms of department, school, faculty and institute, as sometimes people can list those things rather than the university. Limit the search to the last five years and save that list of journals.

2. Now do the same thing, this time searching for prominent researchers based in the same department/faculty as you. You want these peers to be matched as close to your field of research as possible, but not working outside your university. Again, limit the search to the last five years and save that list.

Continued

Activity 9.4: Find Journals You Might Use
– (Continued)

3. The sources that are on both lists are likely to be the places where research can be published that both meets discipline norms AND university standards. Look at the aims and scope of the journals. This will likely come up if you use the words: aims, scope, journal and the journal name. For those with journal names that are also common phrases, the search might be a bit harder. Copy these aims and scopes in a worksheet file and label the key points that tell you that these journals align with your work.

4. In the same document highlight those points that might indicate that your work does not fit. For example, you might be working on drug discovery but your research is for animal health so a human health journal would not work. Or you might be looking at engineering new materials, but need a journal with a focus on composite materials not metals or plastics, *etc.*

5. Determine the most suitable journals and look at the "risks" of not getting your work published in these journals. Where does your work fit best? Which have the lowest risk that a journal editor will email you and say the fit is not right?

6. Talk to your supervisor, former supervisor or a colleague about where they publish their work that is in your area and why. Make a list of at least three key things that experienced people keep in mind and how they look for those things.

7. Assuming you did not already consider all of these journals and techniques from step 6, sit back with your list of journals and review them. Has anything changed? Why or why not? Is there new information about which journals you might publish in?

8. Write a final list of 6 journals that you think are top candidates for your research with a summary of the pros and cons. Put this list somewhere safe and pull it out next time you are looking to plan research. Repeat whenever your goals change.

Takeaway: The place that your research is headed to can (but not always) influence decisions from early in research. For example, you might size your study keeping in mind your research limits and what is considered best practice by the journals you want to publish in. It is best to have a clear view of these considerations while you work, although you should never get overly focused on a particular journal. As research evolves, the right journal will change. As an additional thought, always consider if your research might benefit from some popular media. Helping the public understand research is an important role of academics and it can help develop your communication skills.

Teaching

For an academic research position (as opposed to a job at a national lab or a research institute), progressing your career *via* teaching can be highly rewarding at a personal and professional level. From a personal perspective, you experience the joy of helping someone else make progress. If you have been involved in teaching through your undergraduate or higher degree research training – as a tutor or laboratory supervisor perhaps – you'll know what we mean. If you have not taken those opportunities as yet, we highly recommend that you do when you get the chance. The experiences will help bolster your resume, as well as exposing you to a potentially new area for career progression. Importantly, this helps you boost a range of transferable skills, especially oral and written communication.

As with many other aspects of career development, diversity of experience is extremely useful, so, if you can, take lab tutoring/demonstrating and seminar work, as well as lectures if possible. Each of these different experiences will help better shape your idea of what academic teaching involves and what might be the best area for you to teach in. Or indeed indicate that you are not suited to teaching and instead want to head to a research-focused role at a non-teaching organisation like a national laboratory or a research centre.

If you love teaching and would like to spend much of your career in the room in Figure 9.6, be aware that teaching-only positions at universities

Figure 9.6 Teaching is a central role for most academics, but it uses a different skill set to research. That skill set has the advantage of being very transferrable though. Photo from Sam Balye on Unsplash.

can be regarded below research-teaching roles. It is important to know what the institute sees your role to be and what the culture with regards to teaching is at an individual institution. There are a number of reasons for teaching being viewed (sadly) as less than research at some institutions. The workload can be very heavy, and the career paths are more limited in many places due to many societies putting teaching roles in a personally esteemed but systemically high pressure position. In other cases, the quality of teaching that teaching-only staff provide might be overlooked by those in charge because their interest is focused on research, not teaching. This is particularly likely to happen at research intensive universities. Other universities better balance their dual role of research and teaching. It is important to know to look for this in a potential future employer. If you are interested in a teaching-heavy role, look for the university being very active in education discussions and also career linked discussion in the media, on their websites and at conferences. To work out what universities have a good reputation for teaching performance and have a balanced teaching and research focus, ask a range of colleagues that have taught as academics for a number of year.

Student evaluation of your teaching is an important part of success. Course reviews often have a feedback component. And, if you want to develop as a teacher, seeking evaluations and acting on the outcomes will be an important part of success. The other *evaluation* component of importance is learning outcomes – that is how many students passed the program and how well they performed. Beyond student evaluations, success can also be measured by education research publications if you have an education focus.

Translation/Impact

For those living in a society that believes the market economy will sort out everything (an *economic rationalist* society), the need to justify funding and link it back to real-world outcomes is ever present. As society demands a greater understanding of how taxpayer dollars are spent, the need to closely and directly link funding to impact will be present. Regardless of your position, being aware of the link between your research and real-world impact is important.

I have seen and helped many people make this link. And there are many ways to do it – from licensing to contracts to consulting to partnerships. There are many areas to focus on, not just your widget, your molecule or your idea. For example, I have seen people consult on the establishment of research laboratories. I have seen people license a research method. I have seen people take their analysis and coding skills and apply them to non-research questions (*e.g.*, banking, finance, utility and aged care data). And, of course, I have seen people commercialise their favourite molecules as treatments for diseases, reagents in assays, tools of research and diagnostic markers.

Just like with grants, publication and teaching, the sooner you can start your translation or impact journey, the better. As a higher degree research student, it might be impossible to gain partners. Negotiating is probably the domain of your supervisor, or your university might even say a central unit needs to negotiate the deal(s). There are pros and cons to having third parties negotiate. The pros relate to the art of negotiation and reading contracts (which you can learn about as you go through the process for the first time). The cons relate to the specifics of the deal. You know your work better than anyone. Having others negotiate what you can and cannot do could see them over- or undersell your capabilities.

However, as a student or even a post-doc, your ability to argue against usual protocol is low. So, what can you do to build your career in the area of translation and impact without stepping on the toes of others?

If you haven't even started a project or a section of your project yet, then there might be a chance to direct your work into an area that is very commercially relevant. But for most reading this chapter, that moment has passed. So, the first step you want to take is to understand who your potential partners might be and what they might want. And, given the range of translation and impact options available to you (and mentioned above), it is easy to see how this might be an iterative process. **Who** is interested is probably going to influence **what** they are interested in. And **what** you have is going to influence **who** is interested.

As someone new to this area, the best thing you can do is seek to understand both **who** and **what**. Go out and talk to potential end users (the **who**). Talk to the industry you are part of. The best examples I have seen of this involve students **surveying** the market as they begin their study. They literally seek out potential end users and ask what problems they would like solved. This helps paint a picture of the most useful research the end user might want. This kind of approach also helps build your research skills – you learn about surveys, qualitative and quantitative data collection, data collation, data analysis and visualisation. Bringing that all together into a report that shows your peers or supervisors your decision-making process is also a useful skill. At the end of all of this, if you have the same or similar responses from a large proportion of the end-users you surveyed, you can be relatively confident that addressing that problem will be an idea that will be useful to those partners and perhaps even commercialisable.

Now, if you're reading this book as a(n almost) graduate, you have information and expertise that you can take to **the market now**. You don't need to ask the market. In this instance, you need to share your content with the world and see who is interested. Remember, you need to share more than just your end result. You need to share your journey. The journey helps paint you as a person someone else might be interested in. It gives a more humanistic picture. And like I wrote above, the journey is as valuable as the destination. Look to the social media section below for some more information on getting comfortable about getting active on social media.

Although the number of partnerships can be a good measure of partnership or collaborative success, not all research will lend itself to numerous partnerships or partners. For example, if you are providing a competitive advantage to an organisation, that might prevent you from making a similar partnership with other companies working in the same space. The best example of this would be the provision of a compound or molecule with curative or diagnostic properties for cancer. No company is going to allow you to partner with another pharmaceutical or diagnostic company.

Conversely, if you are performing a quality improvement exercise in the pharmaceutical industry, companies with different focus areas, all interested in, for example, protein purification for different purposes, might be more willing to collaborate. In this example, the companies might not be direct competitors (*e.g.*, food *vs.* pharmaceuticals *vs.* health and safety) and coming together could allow economies of scale or advances to be made that benefit all parties without impacting their ability to compete. Thus, the number of collaborations might be a useful measure of translation impact.

Dollars, too, can indicate success. However, high-cost research can make the collaboration seem more valuable when, in reality, low-cost research might be more profitable. Especially for you, as the researcher. In business, including the business of research, Return on Investment (ROI) takes this into account. If you are able to calculate the cost of your collaboration to your partner (*e.g.* how much they pay you) and present it as a multiple of the income they will receive of money they will save, that would be an ROI. Ideally, all ROIs should be 3:1. That is, for every dollar put into the partnership, business is looking for a three-dollar return or saving. Lower ratios are often not considered because ROI does not always take into account the full cost of the work. In the example above, we did not take into account the staff costs of being involved or the opportunity cost by switching focus to the partnership over other activities.

A survey of your partners and partnerships might be the way to demonstrate value and impact. That is to ask them their thoughts on the partnership. Another useful measure of impact includes repeat collaborations with the same partner, industry or sector.

Of course, talking about translation and impact would not be complete without mentioning patents, non-disclosure or confidentiality agreements (NDAs, CAs) and other ways to formalise intellectual property (IP) protection. However, we will not go into detail about those here, other than to note you should be able to get university support for the protection of IP. You should also review the contract you have with the university to confirm who owns the IP you generate. In many countries, students own IP but staff do not.

Finally, do not discount social media as a way to demonstrate your impact. Yes, you might have started out following my advice, building a content marketing strategy to gain funders. However, if you've now got followers. Lots of followers. Lots relative to similar people or accounts in your field, you're probably making an impact. That, in itself, is worth noting. Beware, not all universities, faculties, schools, departments and institutes see a strong social media following as valuable. So, before you go parading how your content went viral, or you use crowd funding to support your research team, check that these kinds of achievements are considered valuable within your organisation.

Will Extra Training Help Me?

Completing a PhD or master's is a long process. In some countries, it is an expensive one. Then, to talk about NOT going into academia AND re-skilling or up-skilling for a fee seems wrong to Richard (at worst) and unnecessary (at best). Angela has less strong opinions but still generally agrees. Although it should be noted that in countries where there is a very high level of education, it is not uncommon for people to add another qualification or specialty. For example, patent attorneys are often STEM graduates in many countries, not lawyers. This requires more on-the-job training and study. Or if you move into a role with a strong business focus, you might start a diploma of business to better understand this new world. Some then turn this into an MBA. In these examples, these are gained alongside experience. This is key. Post your HDR, you should be looking to focus on experience and occasionally some will decide that that experience should be supported by getting formal qualifications. The experience is paramount not the study.

Thus, if you're keen to develop websites, for example, the best thing you can do is to build some websites. Have some practical examples you can show others. Find websites online and build your version of that site. If you are keen to code, go out and build a game. Or grab some data from public repositories and code the visualisation or analysis of that data. If you want to manage projects, look for examples of projects you have managed in your life and turn them into case studies of your work. If you want to teach, start teaching.

In our experience, doing a short course in web design, coding, project management, communication, social media, training, small business, *etc.*, are all ways to avoid doing the real work. Consider staying away from the realm of formal online (Figure 9.7) or in-person learning, you likely don't need it. The real work is writing a good resume (specific to each job), compiling thoughtful applications to each job and demonstrating/practicing your skills by doing whatever it is.

It might be necessary to have work experience, rather than experience you can guide completely by yourself. You might need to be hands-on with a project or activity in order to demonstrate a skill. It might not be possible to undertake a dummy project. In this case, if you have money set aside for training, use it to fund your experience. How? Well, there are a few approaches. Is there a role you can get that will mean you are adjacent to the work you want to learn about or supporting it in some way? For instance, you might want to learn more about clinical trials. Is there work helping administer a clinical trial where you are a good

Figure 9.7   At this point, you likely know enough, and you are good at learning on the job or teaching yourself. You probably don't need more classes. Photo from Chris Montgomery on Unsplash.

candidate based on your science learning and experience, but maybe you take on a junior role to transition to clinical trials work? Or, if you are in a big organisation, you might volunteer to move temporarily to another group to build a skill when the chance appears. You can also ask your supervisor about internal opportunities to build skills in this manner in a bigger organisation.

If neither of these ideas are possible but you have a budget for training, what about using that money you have saved to take some time between jobs or take leave without pay to shadow a person doing the work you need upskilling in? This is like a professional internship. It will take time and effort to investigate and will not be possible for everyone, but it is a way to improve a specific skill. Remember, you are not trying to become an expert. You are just trying to demonstrate that you have real world experience, that you are willing to get uncomfortable learning through doing, that you are willing to approach strangers to develop a new skill set. These two points will speak volumes to employers, not just about the experience you gain but your planning and forethought will be valued.

Social Media as a Researcher

One of the biggest advantages you have, as a higher degree graduate, is the expertise you gained during that research period.

The internet is full of keyboard warriors, armchair experts, people who have "read the research". Conversely, you actually are an expert. You did the research. You held the pipette. You crunched the numbers. You visualised the data. And it is from this position that you can and should use social media to advance your career.

One of the best places to share your information and journey is on social media – it is free, it is accessible, and lots of people use it every day. Social media has been around long enough that the most established academics will have had social media for the last 10 years of their career. In addition, more recently appointed professors, as well as established researchers, have had social media in their lives since primary school. Yet, there are still MANY academics who feel uncomfortable about self-promotion, and using social media to do so. However, I challenge you to tell me how this kind of promotion is different to a grant, a talk or a journal article? Other than the audience, the idea is the same – get your stuff out there so people think you are good, good enough to fund or collaborate with.

As with the other components of academic success, numbers matter when it comes to translation and impact. The *newness* of social media does provide for some flexibility, but for the same reason (newness) academic administrators also lack awareness of social media and useful metrics of success. So, you might have to learn more about this yourself.

The enemy of done is perfect. I am sure you have heard that saying. In my experience, this is very true for academics and research students who want to use social media. They are so worried about the perfect copy and image that they never post. They look at what their peers or the public are doing on social media and make the decision that their content is not good enough.

The next exercise is to come up with different content ideas to determine what you like to post about and what your audience likes to hear about. The nature of social media means it is very easy to form and find niche audiences. Before social media, if your area of interest was hairs growing on big toes, you'd try to find that society in your local region. Now, you can post about it on social media and quickly find a following. Finding a following much larger than you would have otherwise found. So, don't limit yourself by thinking no one will be interested in your research. Remember, it is an experiment. Content is one of the variables. Don't decide until you have enough data. Some kinds of content you could experiment with include your "hot take" on a news piece or research article; what happened in the lab today; your methods; your reagents; memes about research life; your failed experiments; daily diary; recent successes; where the approaches you use are used in other settings. Anything that formed part of a published poster, article, or presentation can go on social media. Anything that formed part of someone else's poster, article, or presentation can go on social media – academic publishing rules and norms apply, making sure you cite your sources. To practice using social media to help with research promotion, try Activity 9.5.

In addition to this type of content, you should also consider doing opinion or commentary pieces. These will be like literature reviews. But without the painful peer review process at the end. These fit into that extended piece box that you set aside in Activity 9.5.

Together – posting about published research and providing commentary on the state of research in your area of expertise – these posts will help position you as an expert in the field. Particularly in the eyes of the public. Academic researchers have their own hierarchy and expectations around social media. So, don't expect senior academics to always be impressed or to put weight on social media influence. But others will. And it will be a useful tool to demonstrate impact on grants and a low-cost, pyjama-wearing approach to networking.

Activity 9.5: Using Social Media to Help With Promoting Your Research

Activity goal: Practice posting regularly to get comfortable on social media. Repeat later steps once a week to keep visible; several times a week to expand your reach and develop new connections.

Purpose and benefit: Regularly posting and getting used to preparing content will help you understand ways to quickly generate good content and help you build your name. It will help you work out what you like to post and what your audience like to hear about.

Activity steps:

1. Spend a few weeks noting down types of things you might like to post on social media. Start by thinking about anything that you like that other people post or that you would like to hear about. You could also consider your "hot take" on a news piece or research article, what happened in the lab today, your methods, your reagents, memes about research life, your failed experiments, daily diary, recent successes, where the approaches you use are used in other settings. Put aside time a few times a week to go through social media and think about what works and what doesn't. Time limit yourself in doing this to about 20 minutes a maximum of 3 times a week so that you are not over-thinking the notes. Just get a few thoughts and then write notes.

2. At the end of the "thinking and inspiration" phase, sit down and review the notes and summarise your thoughts for ~20 minutes. Make sure to sort ideas into manageable 5–15-minute jobs (e.g., repost an interesting article with your thoughts) and the bigger jobs (e.g., write a blog post on an issue). For now, you are starting small, so focus on the 5–15-minute jobs. You can build to the bigger ones. The larger ones are inspiration for another day.

3. Spend a few minutes thinking about whether there is anything you don't want to post about, or types of posts that you are not interested in. A quick point is when you are on social media, bear in mind who owns the IP and the impact of public disclosure or the ability to use any IP that you might be generating. You can share the general idea of what you are working on without specifics if you are discussing your experiments.

4. Before you actually start writing posts, think about which groups you should connect with in order to be connected with the right people. While you are a student or young professional your number of connections might be smaller than they are just a few years later. But by joining groups in your space, you can get a feed from people and groups you have the most

Continued

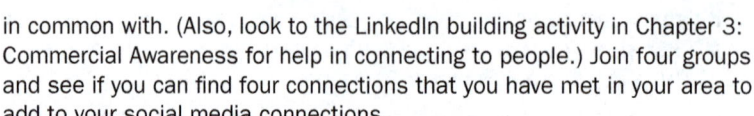

in common with. (Also, look to the LinkedIn building activity in Chapter 3: Commercial Awareness for help in connecting to people.) Join four groups and see if you can find four connections that you have met in your area to add to your social media connections.

5. Many people who are successful in using social media sit down and write a number of posts at once. For example, maybe you are reviewing the literature and you read 12 new papers. Which 4–6 of these would be great to share? When I say great to share, which ones would you like to have had shared with you? Write a couple of sentences for all of these on why you thought each one was insightful, interesting or added to the literature as you finish reading them. This should take 10 minutes per paper only and will probably help you understand the paper better too through reflection. If you want to sit and write about other types of articles instead you can write about popular articles too. Or you might try one of the other types of posts you have identified. The important thing to do is sit and write 4–6 posts using friendly, casual language and have them ready to post. Or if it's a daily activity (*e.g.*, lab highs and lows) set aside 20 minutes every couple of days to write a couple while they are fresh. In this case, try to remember to take photos of your disasters where you can. A broken beaker, a flat line on your instrument, a flooded river or an empty shelf waiting for supplies. None of these are artworks, they just help to support your story.

6. Now every 2–3 days go onto your account and share an article with your thoughts. It is useful to use a photo when you post but do not overthink it, it just has to be loosely associated with the topic. A post about a medicine might just have a picture of healthy people being active, one about microbes might have a microscopy image of a common microbe, *etc.* If you are worried you need to do many posts together, you can have them mocked up already and just copy and paste so you literally just need 1 minute of your lunch break. Some platforms also let you make posts beforehand and schedule the release.

7. Try this approach with 2–3 different types of posts each time for a week or two. What did you learn in this time? What sort of posts did you like putting up and why? Were there types of posts that got more likes and views? Were you able to produce posts in a shorter amount of time? Did you get your time down over the weeks? What approach(es) might you take forward? Remember, you don't have to just use one approach, it is normal to mix them.

8. Revisit your contacts and links to organisations and businesses now that you have had a few weeks of seeing what works, what's online and who is

Activity 9.5: Using Social Media to Help With Promoting Your Research – (*Continued*)

posting what. Are there new connections to add? Did you get hits on your posts and then connect to people? This is a great idea where you get the chance.

9. Write a plan of what you want to get out of social media in the next six months.

Takeaway: There is no right or wrong level of social media interaction. It might be that you decide it's not for you. At least you know and understand social media a lot more now. Alternatively, hopefully you have gained a few social media skills that you can choose to use when you want, or maybe you have been bitten by the social media bug and want to learn more. There are large numbers of free learning materials about how to use social media platforms, including a range on LinkedIn Learning.

Are You Still Grappling With the Way Forward?

For most people, change is hard. Without a doubt, that is the case for Richard. Richard stayed at university because it was the easier, no change, thing to do. Rather than graduate and start a career, Richard chose to do honours. Rather than finish honours and look for work, he chose to do a PhD. It was only at the transition from PhD to post-doc that he started making active choices about his career. And, if he's honest, it was only because choosing further study was not an option. Others might stay on an academic path due to family expectations. Both of these processes are understandable. No judgement here. Both these circumstances mean you got to your higher degree, and you haven't put a lot of active thought into why you want to be there, or what you want your future career to look like and why.

That brings us to graduating your higher degree and making a decision to leave the research or stay within it. You can read a short summary of the decision process Richard went through in Box 9.1. Regardless of the choice you make, there are two things to keep in mind:

1. You can always change your decision, and progress along different path.
2. Making a conscious choice is better than drifting to the easiest thing.

	Box 9.1 Leaving Academia

Richard left academia for two main reasons:

- My perception of **the grant treadmill**. Specifically, I felt like research was essentially a never-ending loop of write grant → review grant → submit grant → get funded/rejected → write grant… I did not want that kind of lifestyle for myself. Not to mention the insecurity of living from grant to grant, contract to contract.
- I also felt there was a **disconnection between effort and success.** In the research group I was part of, I saw people put in lots of hours, keep up with reading the literature, have good experimental design and technique, only to have results pan out in an unpublishable way. I also saw the opposite. Low hours. Less literature and experimental understanding, only to see those people have publishable research findings. Again, I did not want to work in an environment where more effort did not lead to more or better outcomes.

So, I decided to leave.

And eventually set up my own business.

And there I was on soft money. Worse than grants. Shorter term. No security. Less funding. BUT the key was the enjoyment of the work. So, I put up with the ambiguity of being self-employed. And, in small business, there definitely is a strong link between effort and outcome, so I was happier that I could see my hard work paying off.

So, when thinking about why you might leave academia, bear in mind it could be the type of work as much as the setting.

What would bring me back? I was (and still am) enthralled by the idea of discovery. I loved being the first person to find something out. The first person, perhaps in the world, but certainly amongst a very small group of people to know something about how cells work. Even as I write this, that knowledge thrills me.

Have Some Conversations

In order to think through what you want to do, you need to learn more about roles you might go into and what those roles might entail. We have started you off above by laying out what different jobs tend to entail. Of course, it is always wise to talk to those that you can in your specific area (both geographical and discipline) to get extra details on the type of positions and the pros and cons in your exact environment. Just like any other student, the careers office and your former teachers or current supervisor are good places to start conversations to learn more (unless you are under pressure from that person to make a particular decision).

Figure 9.8 To help you decide what you want to do, get out there and have conversations to work out what is next. Photo from Tim Gouw on Unsplash.

Hopefully, you have some connections that are currently in the work-force. And hopefully you look a little less worried than this person in Figure 9.8, but if you are worried, that's OK. The more opinions the better, but two opinions are definitely better than none, so talking to anyone is valuable. Use your connections to find people to fill your bias gap(s). Consult the networking chapter (Chapter 4) if you want help with this. What do you not know anything about? Maybe you need to broaden out your connections on LinkedIn to help you talk to people. This might be connecting with people you already know, or it might be making new connections, for example, when you see someone often posts the type of content you like it's good to reach out, thank them for their content and ask if they would like to connect. Think outside the box. Chat with your barista or local restaurant owner about how they ended up working in a café, talk to a wide range of people about what they love or do not like about their careers and see if you learn anything that can help you.

Outside of university or discipline connections, if you are still having trouble, or just want to have a more thorough investigation, the next group of people to talk to are relatives and family friends. In my experi-ence, a lot of people overlook relatives as a source of useful information. Setting out to have the specific discussion is important since there are so many other things you could talk about. Make sure you take into account their specific position and where they are coming from as well. Have they had a lot of career success? How does the area they went into dif-fer from yours and how has it changed since they started work? Consult with your aunt, the one who is always taking risks herself. Chat with your uncle, the one who is always taking a counterpoint to your dad. Great career insights are still great career insights if they come from someone from a prior generation or another field. You just have to be good at working out what is an insight and what is not of use.

In all cases above, these conversations should be about getting a complete picture. So, try to understand what day-to-day work life is like. Try to understand what drove a particular decision – *e.g.,* moving to a new city, state or country. Get to know why they DIDN'T do something as well as why they DID other things. Get to know what made certain decisions more or less risky or safe and what they did to make risky decisions **safer**.

Head or Heart?

By now, you probably have a wealth of information about academic *vs.* a range of non-academic workplaces and roles. You also have hopefully learned from all the other chapters in the book which roles are suitable for those that have done HDR and those that have not. Hopefully, you have even been able to work towards some much narrower ideas about what you want out of your future work and why. But now there is an extra question: are you going to let your head lead or your heart? Both can have a vote, but does one dominate?

Think with your head might mean making a spreadsheet, with a list of factors. Each factor would have a weighting for how important it is for you. Then, based on your conversations and readings about careers over a period of time, all factors get a score. Then you might multiply the score by the weighting. So, academia might get a score of 3 for job security (out of ten), and you might weigh that as a 4 (out of 10) for importance. Giving a score of 7 out of 20 on that factor. Then you repeat for other factors – pay, lifestyle, flexibility – whatever you feel need to be included in the analysis. There is no right or wrong way to do this, but it is an approach.

Alternatively, deciding with your heart could include discussion with a trusted friend or partner (or several) to see how you feel about different aspects of future work. Of course, you need to bear in mind this discussion is kind of the opposite of the information gathering exercises above. This time you can sit down with someone close and they can ask you about your passion, what you are worried about and why, what are the emotional "must haves", *etc.* Really, these people are sounding boards. We know other people who encourage talking to yourself. Recording yourself or writing an essay to yourself or writing a reflection as per Chapter 2: Transferable Skills and Reflection. They are all possibilities too. As is drawing (literally) how all of the different factors relate; or drawing how each of the different roles or careers or sectors makes you feel. Although scientific education has probably taught you to be analytical, taking a more creative or intuitive approach might highlight factors that were not clear. Conversely, it might make the analytic choice even more obvious.

Knowing your preferences for head or heart-based decision making can help you make a choice, but it isn't one or the other. You can, of course, go through both internal processes and listen to both. Most people will make decisions based on both head and heart dimensions.

Check Your Progress

In the first chapter you likely filled in a short survey to gauge your perception of your employability, which will have been emailed to you. This was called a self-perceived employability questionnaire (Activity 1.1). The subject on the email will have the title of the book in it. If you are ready to check your progress, use this QR code to recomplete the questionnaire and it will send the new results. Compare these results with any you have been sent previously by doing the following:

1. Bring the results up together and compare your results. If you are later in the book, you might have more than two. You can compare which results you want, or even all of them. Transferring the data to an Excel spreadsheet will help if you have many answers and are inclined to do that, or if you need an activity to help develop your Excel skills.
2. What are the differences? Sometimes you will feel a bit different on a different day, so not every value that changes by one point is a significant change. Maybe you had a good day or a bad day. Can you see any trends or big differences? Do any of the changes feel particularly true? Even if it is just one point.
3. Often when people learn more about a subject, they first loose a bit of confidence as they are starting to learn about how much there is to learn. Has that happened to you? This is an important process to be aware of. Maybe you aren't feeling as comfortable with your employability but that's because you had to realise how much there was to know in order to be able to go out and learn it. Well done, that was an important realisation.
4. Now that you have stood back and thought about your overall perceptions of your employability, are there any plans you think you might make or change? Or maybe something you have realised you want to prioritise or deprioritise? If so, make a note of these and set a calendar reminder once a week for the next month to ensure it doesn't get forgotten.
5. Enjoy the next chapter!

Developing Your STEM Career: What Next?

ANGELA ZIEBELL

Deakin University School of Life and Environmental
Science, Australia

So, you have made it to the last chapter. Maybe not directly, but there is no right way to travel through this book. Some things are hopefully clearer. While you might have gotten uncomfortable at some points while challenging yourself to work through the activities, hopefully there has been a gain in your career understanding. A few of you who weren't thinking much about a career path before picking up this book might have had a bit of a fright when they started to learn how much there can be to think about. But this is a positive. You now understand how much you can learn and influence your own STEM professional development.

Before I sign off and go back to teaching chemistry, we have a couple more activities for you. First, let's reflect on what you learnt and how you learnt it in order to evaluate and articulate how much you learnt (Figure 10.1). Remembering back to Chapter 1: Introduction, there was an activity that had you reflect on how you felt about your self-perceived employability. Now we are going to do that activity again so that you can reflect on those questions, but also what the change since Chapter 1 means to you (the original results will have been emailed to you if you filled in the email address). This time we have added a couple more short-answer questions to extend your thinking. Work through Activity 10.1 to see how your journey through the book has impacted your perceptions of your employability. The book map is included for reference (Figure 10.1).

> *"If you can't say what you want, and make a plan for getting there, any progress will be a coincidence."*
>
> – Angela Ziebell.

Figure 10.1   Map of this book.

Activity 10.1: What Are Your Self-perceptions About Your Employability Now?

Activity goal: Explore your self-perceptions around your employability and check how it has changed from the start of the book.

Purpose and benefit: In order to understand how your thinking about your own employability has changed and where you are at now, complete this series

Continued

Activity 10.1: What Are Your Self-perceptions About Your Employability Now? – (*Continued*)

of questions, which research has shown is a reliable insight into your perception of your employability. You will then compare with the result you got when you first did the Chapter 1 version of this Activity (Activity 1.1) and reflect on the differences or maybe the lack of differences in some places.

Activity steps:

1. Follow the QR code above to the questions. Complete them and get a copy sent to your email for reference (or if you just completed the Chapter 9 questionnaire you can use that as it is the same).

2. Review the results you were emailed (they will have the title of the book in the subject line) and compare them to the original, which you were sent when you first read Chapter 1 and did Activity 1.1. Think about what you answered in the additional Activity 10.1 questions if it seems relevant.

3. In Activity 1.1, you were also directed to save your reflections and notes to a file to use throughout the book. If you have these notes, dust them off and see what they add to your understanding of how your self-perception of your employability has developed.

4. Write out your thoughts. Are there any surprises? Jot down a few thoughts about either the questions or your responses. If you have lost the email or suspect you didn't complete Activity 1.1 this activity will be harder, but you might recall some differences. Look through the questionnaire and think about each item. What do you think you would have answered before you read this book and worked through the activities?

5. Take either a point that you were surprised about or a point that you fully understand but want to work on articulating better. Write 3 thoughts related to the point you have chosen. Reflect on what the change or lack of change means to you and why it has come about. Use the Gibbs reflective cycle from Chapter 2: Transferable Skills and Reflection. Does it show progress? Does it indicate you haven't gotten as far as you hoped? If it's a surprise, what does it mean? How did it come about? There aren't any wrong answers. This process is usually helped by not over-thinking the process or your answers.

6. Once you have reflected on your self-perception of your employability, think about what that means for where you are headed in the next 12 months. Are you better positioned? If not, how can you work to help yourself

> ### Activity 10.1: What Are Your Self-perceptions About Your Employability Now? – (*Continued*)
>
>
>
> be in a better position? What in the book can you go back to in order to access help? Make a plan for what to do over the next three months, including fortnightly check-ins with the plan to see if you are making progress.
>
> **Takeaway:** Better understand how far you have come in your employability journey. Most people will make quite a bit of progress, but if you haven't, it's important to take stock and work out why. Maybe something came up in your life this year that made progress hard. Maybe you have increased your understanding of how complicated career development can be, so you feel like you went backwards in self-perception before starting to move forward again (not at all uncommon). The most important thing is to understand and be conscious of what is happening so you can work on it.

What Now?

What is next is going to depend a lot on what you want and where you are on your journey. So, let's look at a few different career journey stages where people can be at this point and then look at one last activity.

Students in the First Two-thirds of Their Degree

If you are at this point in your journey, you have a lot of time to refine what you want from a career at the moment, but that time will go fast. Remember, because you are starting your career thinking early in your studies, over time what you want will change rapidly for many students as they learn more about possible future workplaces. Returning to repeat the activities in the book as you refine your career thinking and come across new ideas is an excellent habit to get into. Doing so doesn't have to take a lot of time, maybe an hour while commuting every few weeks. Students often find that some things change a lot, many things are just being fine-tuned as you work through what you want, and some career goals or plans stay central to what you want. But until you start working on exactly what you want from your career, it is hard to know what will change and evolve, and what will not.

If you haven't had a professional adjacent role yet, look for opportunities like internships (preferably paid), or professional adjacent roles (think administrative assistant in a laboratory, optometrist, or medical office; data entry, cleaning or management; laboratory or engineering assistant; IT helpdesk or junior programmer/developer). There are a lot of casual and part-time jobs that fit this bill; however, it isn't unusual for students to stay in retail and service jobs throughout university. If you look for professional adjacent roles, your experience and workplace knowledge can give you a competitive advantage on graduation. Experience that resembles more professional work (especially a role you are looking to move into) will look great on your graduate resume and give you professional references to call on when applying for work post-graduation.

It is also a good idea to keep in mind what transferrable skills and knowledge your current workforce experiences have helped you develop. If you have developed as much as you will in a position and have a good track record, it might be time to move on to something else that expands your skills/knowledge. Look for roles that develop transferrable skills you need to fine tune for likely post-graduation roles. This could be because you personally feel the skills are weaker and need to be strengthened, or because that particular skill is in demand in many post-graduation workplaces you are considering.

At the end of the day, great work for getting to the end of the book pre-graduation. Starting early will kick start your entry into the workforce post-graduation. So, get that pen and paper out like this student in Figure 10.2, or make an amazing spreadsheet and make a plan, proof it with others and understand that plans evolve, and that's normal.

Figure 10.2 Planning ahead is a student's best career tool. Work out one thing to achieve each semester and you will accrue new skills and insights all through your degree. Photo from Gabrielle Henderson on Unsplash.

Students in the Last Third of Their Degree

If you are in the last third of your degree, this is a time when you have likely got a lot of enthusiasm for getting a job that works for you in the next year to 18 months. Try to use the tasks in the book to move forward one thought or process at time and not to get too overwhelmed (I understand this can be difficult). Having a to-do list and keeping track of progress is a simple but rewarding way to see your progress. Make each task you list down manageable. For example, don't put down: write new CV. You could break this into: find referees, bullet point out full plan, write in details, ask friend to review, research and pick a format. As one task, it really is overwhelming to think about writing a whole professional CV, but it's not one task. There are lots of elements that go into a CV and focusing on one at a time helps a lot.

Enthusiasm for getting a new or first professional job might tip into anxiety. That's normal, but remember, you have an advantage over many students. Even getting as far as reading a few of the chapters in this book and doing the activities will have helped you understand your career and the processing surrounding it a lot better. With this book, you also have a guide to get you through putting together new job applications and preparing for interviews. While we understand that these things take a lot of effort and tend to make people nervous, using this book to help you can minimise wasted effort.

Professionals Upskilling

Many people get through part of their early career without having a particular plan or learning a lot about how to develop as a professional. I was like that. I wanted to work in chemistry research, and I lived near a big national research facility where I had done a research project while I studied. I knew enough to know that it helps to know professionals before you leave the university, and that employers want to see that you have work experience that shows you have proven your work ethic (*e.g.*, hard-working, punctual, trustworthy), usually in a casual job. But that was about it. I was even told by my second employer a while after I was hired that I was given the 12-month contract, and another young graduate got the ongoing position because I didn't interview anywhere near as well. This led to me having to compete for another position towards the end of my 12-month term. This could have been avoided had I known how to prepare for an interview.

If this sounds like you, you have ended up in the right place by finding a resource to help you. Through working, you have a solid understanding of being an employee and you will have been able to see which pieces of advice can help you and how, in a very real and memorable way. Hopefully, you have been able to use some of what is in the book already as you were reading because you will have your work life unfolding around you and have great examples to reflect on, and skills that you were developing.

But now what? I actually think many of you here will already have an answer for this question, because as you worked through the book you would have been refining your understanding of your transferrable skills, dusting off your CV, advancing your interview and networking skills and slowly upgrading yourself as an employee. If you are honest with yourself, I think you will see there have already been quite a few changes to your understanding of the workforce and where you might go in it.

However, most of you will still be early in your career, so there is no need to rush and there will naturally be changes in what you want over time. So, coming back to the practice of reflection, keeping your CV up to date, keeping up with your workplaces skills, and following any plans you made to add skills. It is important to work on these tasks slowly so that you are fully prepared for whatever comes next.

Insights From Others

With those thoughts in mind, let's hear one last time from each of our authors (in no particular order). I asked every author to share with us their top tips and insights into the world of work and career development. In your career journey, you will probably find you relate to some professionals and mentors much more than others, so I thought it was best to hear from all of us.

Dr Rebecca Yee – Energy systems engineering and STEM communications advocate

Working on this book has forced me to reflect on my own career shaping efforts. I would even let on that this is very much a case of 'do what I say, not what I've done'. While I have enjoyed a lot of my jobs and feel fortunate in many ways, there is always more that could have been done. I can recall many interviews (even recently) where I have not been sufficiently prepared (do not underestimate the STAR approach). I have often felt overwhelmed when looking for a job and remember being extremely stressed as a graduate, juggling final exams and casual work while feeling pressure to get a 'real job'. To this I would advise the added transferrable skill of *patience*. Look after yourself. Identify your own priorities.

A couple of decades on, and with more understanding of economics, my main takeaways are thus:

1. **There is a *HUGE* demand for STEM skills**, in STEM fields or companies themselves, but really every organization that deals with numbers in any way boils down to STEM (this is nearly everyone). Rest
   assured that there is something out there for you. It just may not have
   your degree or skills list in the job title! You will learn a lot just by
   looking around even, especially, when you are not under immediate
   pressure to get a job straight away. Explore *industries* and *companies*,
   not just the scientific topics that take your fancy. Try to figure out who
   does what and who owns what.
2. **There are trade-offs to be made.** With most of our lifestyle driven by
   immediate gratification, it can feel like you need to tick every box right
   away. For example, like many of your ancestors, you may need to move
   to where the grass is greener. The world is still a physical place and while
   the pandemic and digitisation have allowed us to connect anywhere, you
   may still need to move to align your skills and interests to where the work
   is. This can be hard but, in my opinion, so worthwhile. You may also need
   to take on jobs that aren't a perfect fit (is there such a thing?) or may just
   take on a job that you have no idea about just so you can learn what it is!
3. **Things change.** My priorities now are much different than 10 years
   ago. Reflecting and identifying these values has really helped me to
   feel pretty balanced about my work and life. There have definitely
   been times where the seesaw has tipped in either direction, but that
   is life itself after all.
4. **Find something to care about.** This can, of course, be outside of your
   job or career, but if you can at least identify your own values, this will
   help give you a sense of purpose. Feeling like you are working towards
   something bigger than yourself is a valuable strength.
5. **Connect with your community.** This is tied to the above, but with
   more emphasis on the fact that we are social beings. *You* are an extension of your society. And keep in mind that while *you* may have several years of STEM education and experience, *most* people do not.
   Share what you have learned! Celebrate and integrate the role of science and engineering in society and in general conversation. This is
   why we harp on about networking, not just as a 'transferrable skill
   useful for your career', but as a critical aspect of what makes you a
   human being.
6. **You don't know what you don't know.** Step outside your comfort zone.

Coming back to my own reflections again, I can even share that when
asked to edit this book (by someone I had previously tutored with, aka

my network), I had little idea of what such an editing job would even entail!

> *"I have learned a lot by doing."*

Dr Sophie Mckenzie – IT academic with a passion for supporting students to achieve their career goals through practice-oriented experiences and assessment

> *"Building a career portfolio is an ongoing and evolving activity."*

Your career portfolio is a tool to help you showcase your current skills, goals, and plans to get a job. But the process of creating your career portfolio is just as important. When building your career portfolio, you will collect numerous artifacts to demonstrate your skills and abilities, but more importantly, you will engage in self-reflection to consider your preparation for work. Building a portfolio during your studies is not just about collecting cool artifacts; it is a tool for reflecting on your growth and getting ready for an amazing career. You can use your career portfolio in many other parts of the career building process, such as networking, in your job application and for building your online profile.

It is important to emphasise that building a career portfolio is an ongoing and evolving activity. Your career portfolio is a constant work in progress because each year brings new stuff – skills, experiences, and interests. You will regularly revisit and refine your portfolio based on your current career interests. However, building your career portfolio does not have to be a big job! You can start by simply saving and recording in one place things that evidence your skills and achievements, such as assessments or awards. Once you have a few things to get started, you can reflect on what these artifacts are saying about your skills and career goals.

When you are ready to formalise your career portfolio, it will have sections that help you demonstrate, describe, and reflect upon your professional self and your achievements. Your portfolio will highlight to employers your broader career goals, while also showing/evidencing your skills. It is up to you to decide what sections to include and how much you would like to share with others. A career portfolio may take several different formats, but will describe yourself, your career goals, provide a professional summary (in relation to jobs in your chosen discipline) and showcase your skills. One benefit of using a career portfolio is that you can tailor and present your career information and professional self in a format that best appeals to you and your unique skills and abilities. In fact, making yourself 'stand out from the crowd' can be achieved through a portfolio, as you can tailor content presentation and style.

Your career portfolio should also make you adventurous about career exploration and building a career portfolio, suggesting that considering multiple possibilities and learning experiences can lead to a fulfilling role once you have finished studying. Ultimately, curating a portfolio during studies is a valuable tool for self-reflection, better articulation of personal and professional growth, and preparation for a successful career.

Dimantha Harshapriya – An early career researcher working on marine hydrodynamics, who loves math, ships, writing and dogs

"Preparing a toolkit without knowing the specifics of the task you're destined for."

In the dynamic tapestry of your career, every thread counts, and nothing you learn is ever wasted. It's akin to preparing a toolkit without knowing the specifics of the task you're destined for. Throughout my own journey, there were instances when seemingly unrelated skills or knowledge unexpectedly emerged as invaluable assets. One episode that comes to mind is my early foray into computer programming in several languages. Initially perceived as a sidetrack from my core focus, those coding skills became my navigational tool in complex simulations. Fast forward to a position as an early researcher at the world's top department in my field. It turns out that understanding algorithms opened doors to many innovative solutions. The lesson? Embrace diverse experiences because they mould you into a versatile professional. It's like planting seeds without knowing which ones will blossom first. You might find yourself marvelling at the resilience you developed during a challenging group project, or leveraging a coding language you once considered a mere elective. The beauty lies in the unpredictability; your career is a mosaic, not a linear path. So, as you navigate the twists and turns, remember that every bit of knowledge is a stepping stone, and you never know when the skills you've acquired will become the linchpin in your success story. Stay curious, stay adaptable, and trust that your unique blend of experiences will craft a narrative that is uniquely yours.

Dr Angela Ziebell – Chemistry/STEM academic interested in designing learning to prepare students to contribute positively to the world

"A path not chosen is just that. It does not need to be a regret."

I've always been very conscious of the environment and humanity's role in how the world is changing. When I was in high school, I was interested in becoming an environmental scientist. I didn't really know a lot about what exactly an environmental scientist did (my first mistake) but it seemed like they must use science in different ways to look after the environment and that sounded like a good job for me. My parents convinced me that huge numbers of people would have the same thought and would also want to work as environmental scientists and therefore it would be hard to get a job. Plus, no one pays to look after the environment that well, so pay will be terrible (this was Australia in the 90s, but it applies to other places and more recent times too). While there were definitely some limitations on how good their advice was, their points seemed logical, so I made a strategic decision to look at something different as a career. I loved chemistry, so I became a chemist (despite my year 12 chemistry teacher making it clear he thought I should move out of chemistry). I specialised in green and sustainable chemistry in my studies, although half of my professional work has not been in that area.

Fast forward the clock to today and there are a lot of environmental scientists in most countries, and there are career paths with options to work for the government or in the corporate world, or in research. Do I regret the decision to change to a new path? No. I don't believe there is one set path. I am currently working on how we as a university, but also how we as a country, embed climate change education into all university training so that all graduates have an understanding of how climate change will be relevant in their job. Think engineers choosing the more environmentally friendly materials, carbon accounting for accountants, how to teach around climate anxiety for teachers, *etc.* I would have never imagined when I was in high school that I could work on this. So, while imagination in your career is an amazing thing, some paths just aren't obvious yet and a path not chosen is just that. It does need to be a regret.

Pick the best path that you can for that time and keep yourself updated as both you and the world around you change. You have the aid of this book to help you work out what you want, and work out who can help you consider, build on, and refine your goals. So, with attention to the advice in the book and lots of preparation, hopefully you are more aware when making decisions.

Making you independent career developers is our real aim in this book. There is no one right way forward for anyone at any time. Sometimes the number of options is the problem! No experience or skill will ever truly go to waste, you just don't know what your path looks like yet. While I understand this can cause some trepidation, for most people, I think that this is a very feature of moving forward in your career.

Rose Herbert – Molecular biologist and faculty career skills lead

'Never underestimate the power of telling other people about the fact you need a job.'

My own career path has been relatively boring – I've always been passionate about science and teaching from when I was very young (although initially I had my heart set on veterinary medicine). Instead, I have two stories in particular that I'd like to share with you from my time helping graduates find and apply for roles that interest them.

Never underestimate the power of telling other people about the fact you need a job! My brother-in-law was looking for a job after he had graduated from his bachelor's degree. He had only applied for a handful of 'ideal jobs' and hadn't put much effort into his job application documents. Someone that I'd previously met at a LEGO convention happened to text me and ask if I knew anyone who was looking for a job. I asked my brother-in-law if he'd be open to the job and the next thing I knew, he was being prepared for the interview. Three years down the line, he is still happily employed and moving sideways into an area of economics that he is trained in, and passionate about.

I've also worked with a PhD graduate who had been applying for five jobs a day, five days a week for the last year after completing their degree. Despite this huge volume of applications, she was yet to land a job interview. When we sat down together to brainstorm improvements to her resume, it turned out that she had a lot of volunteering experience that simply wasn't represented there. We tweaked the resume and did an overhaul of her cover letter to better showcase her project management skills rather than research skills. This was sufficient to be offered a casual entry-level role in a clinical trials organisation, and since then she has been promoted up the ladder to a senior position.

Something that most excited me about contributing to this book was the chance to impact more science graduates in finding a career that they are passionate about. I work with around 800 students a year, but this is so few compared to the number of scientists we need for the future.

Everyone deserves an opportunity to learn about how to make the most of their talents and interests. If you are going to spend one third of your life at work, then that work should be meaningful.

The Last Step Is to Continue

You have likely put in many, many hours working through this book. Now that you have also reflected on your self-perception of your employability, the main focus needs to be for you to keep putting time into developing yourself. While setting aside time is difficult for most of us, the work you put into determining what you want in your career, how you can develop yourself as a professional, and what you want out of a career is truly priceless.

I strongly encourage you to find a time of the week or month that you can generally get an hour or two to focus. Maybe this is on public transport or in the morning on the weekend. If you schedule this time, the first part of the career development challenge is taken care of – time. Then use that time to work on whatever it is that you have decided will help you progress. Don't try to do it all at once. You can have a focus for 3–6 months and then, when you feel like you have made good progress (*e.g.*, your commercialisation skills have improved noticeably, or you are getting used to networking), move on to your next career development point.

If you are spending a lot of time applying for jobs (for this you will need more than 1–2 hours a week), then developing great CV and cover letter writing skills or practicing interview questions is probably what you will focus on. This is a lot of work; you will get overwhelmed if you add much else. For all things career development, time spent slowly adding skills will surely add up over time and you will look back and see the progress you have made.

"Every career path is different because every individual is different. It really is just that simple."

– Angela Ziebell.

The whole team wishes you the best of luck with your STEM career, and we can't wait to have our book on the shelves helping hard-working STEM students and graduates.

Index